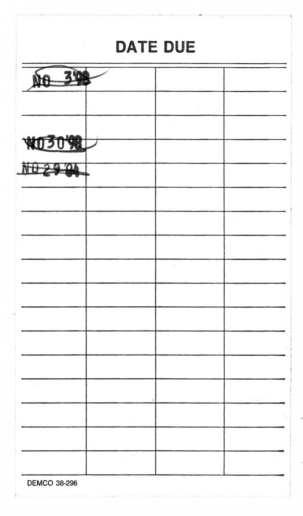

DATE DUE

NO 3'98			
NO 30'98			
NO 29 04			

DEMCO 38-296

Cultures in Conflict

Cultures in Conflict

CHRISTIANS, MUSLIMS, AND JEWS
IN THE AGE OF DISCOVERY

Bernard Lewis

New York Oxford
OXFORD UNIVERSITY PRESS
1995

Oxford University Press

Oxford   New York   Toronto
Delhi   Bombay   Calcutta   Madras   Karachi
Kuala Lumpur   Singapore   Hong Kong   Tokyo
Nairobi   Dar es Salaam   Cape Town
Melbourne   Auckland   Madrid

and associated companies in
Berlin   Ibadan

Copyright © 1995 by Oxford University Press, Inc.

Published by Oxford University Press, Inc.
200 Madison Avenue, New York, New York 10016

Oxford is a registered trademark of Oxford University Press

Library of Congress Cataloging-in-Publication Data
Lewis, Bernard.
Cultures in conflict : Christians, Muslims, and Jews
in the age of discovery / Bernard Lewis.
p. cm.   Based on the Merle Curti lectures delivered
at the University of Wisconsin–Madison, May 1993.
Includes bibliographical references and index.
ISBN 0–19–509026–8
1. Europe—History—1492–1517.
2. Europe—History—15th century.
3. Europe—Territorial expansion.
4. Jews—History—70–1789.
5. Islamic Empire—History—1258–1517.
I. Title.   D228.L49   1995
940.2′.1—dc20   94–14468

2 4 6 8 9 7 5 3 1

Printed in the United States of America
on acid-free paper

Preface

*T*HIS BOOK is based on the Merle Curti Lectures delivered at the University of Wisconsin, Madison, in May 1993. My theme in these lectures was the multiple anniversaries of 1492 and the sharply contrasting ways in which they had been perceived and commemorated in different places and among different groups of people. My purpose was to draw attention to some of the other events of that year, notably the Christian conquest of Granada, the last outpost of Muslim power in the Iberian Peninsula, and the expulsion a few months later of the Jews from all of Spain; to show how these apparently disparate events were closely interrelated; and, in so doing, to situate the three processes—of conquest, expulsion, and discovery—in a larger context of international, interreligious, and, one might even say, intercontinental history.

The Christian capture of Granada marked the completion of the long struggle for the recovery and reconquest of western Europe and prepared the ground for the great European counterattack, of which the voyages of the Spanish and Portuguese navigators were, in a profound sense, a beginning.

By 1492, Christian rule prevailed everywhere in the Iberian Peninsula, but many non-Christians—Jews and, in far greater numbers, Muslims—remained. Their removal was seen by the Christian rulers of Spain and, a few years later, of Portugal as a necessary part of the process of reconquest and counterattack. The expulsion of the Jews, the smaller and weaker of the two minorities, was the first step. The expulsion of the Muslims—a larger, more complex, and more dangerous task—took somewhat longer.

The expansion of Europe took place simultaneously at both ends, and in both it began with the effort to remove a Muslim domination that had lasted for centuries. In the east, the Russians, after a long struggle, were able finally to throw off what their historians call "the Tatar yoke." In the west, the Christians put an end to the Moorish domination established eight centuries previously and, in time, even to a Moorish presence. In both eastern and western Europe, the resurgent Christians carried the struggle into the enemy camp. In the east, the Russians pursued the Tatars into Tartary and established a domination over Central and northern Asia. In the west, the Spaniards and Portuguese, followed by the other maritime peoples, pursued the Moors into Africa and Asia and, almost incidentally, "discovered" and colonized America.

In preparing this version, I have drawn on material previously included in the Bradley Lecture at the American Heritage

Foundation in Washington, D.C., the Kalman Lassner Lecture at Tel Aviv University, a public lecture at the New York Public Library, and guest lectures given under the auspices of commemorative committees in Genoa and Istanbul, all in 1992. My thanks are due to my hosts on all these occasions and to my various audiences, whose questions enabled me to sharpen my thoughts and clarify their exposition.

My thanks are due to a number of authorities for permission to reproduce illustrations: for illustrations 1 and 2, the Ministry of Culture of the Turkish Republic; for illustration 3, the Réunion des Musées Nationaux de France; for 4, the Alfred Rubens Collection in London; for 6 and 7, the L. A. Mayer Memorial Institute for Islamic Art in Jerusalem; and for 8, the Trustees of the British Museum.

It is also my pleasant duty to thank Professor Nurhan Atasoy of Istanbul University and Mariëlla Beuker of the Jewish Historical Museum in Amstedam for help in locating and obtaining illustrations; Jane Baun and David Marmer, both of Princeton University, for much valuable help in the preparation of my manuscript; and Nancy Lane and Irene Pavitt of Oxford University Press for their help and advice at various stages of publication.

Princeton B. L.
February 1994

Contents

Chapter One
Conquest, 3

Chapter Two
Expulsion, 27

Chapter Three
Discovery, 55

Notes, 81

Index, 95

Illustrations

Ilustrations follow page 58

1. Part of Piri Reis's world map, showing the New World

2. Christopher Columbus at the court of King Ferdinand

3. Portuguese Jew

4. North African Jewish peddler

5. European visitor to Iran

6. Jesuit priest visiting an elderly Indian ruler

7. Young Portuguese woman

8. Woman of the New World

Cultures in Conflict

CHAPTER 1
Conquest

*F*OR MOST AMERICANS and Europeans, as well as many others who learned history from American or European teachers or textbooks, 1492 was chiefly memorable as the year in which Columbus discovered America. In some countries and in some communities, other events that took place in that year were celebrated or commemorated. But even those concerned with such other commemorations would surely have agreed, until not so long ago, that of all the events of that crowded year, the discovery of America was by far the most momentous in the history of the world and in the history of humanity. As the quincentennial year 1992 approached, elaborate preparations were made—in Genoa, where Columbus was born; in Spain, from which he set sail; and in all the American lands whose modern history customarily begins with his arrival.

But as the preparations to celebrate the discovery progressed, discordant voices were heard, casting doubt on the entire basis of the intended festivities. The word *discovery*, it was explained, is both misleading and demeaning. The Western Hemisphere was, after all, there before the white man arrived to conquer and exploit it, and his arrival brought not so much the creation of a new civilization as the destruction of several old ones. The term *America* was inappropriate to a continent that had flourished long before an Italian cartographer called Amerigo Vespucci made the momentous discovery that the place on the far side of Atlantic was not Asia, as Europeans had thought, but somewhere else, and was rewarded for this discovery by having his given name, in a somewhat garbled form, splashed across the map of the Western Hemisphere. And as for celebration, it was made perfectly clear that this was an occasion not for celebration, but for lamentation on the part of the victims and for penitence and self-flagellation on the part of those who could in any way be identified with the perpetrators.

Even Columbus himself, the erstwhile and intended hero of the occasion, did not escape historical revisionism, and in the reassessments of his character and motives, he appears at times as a combination of the less amiable features of Attila the Hun, Jengiz Khan, and Adolf Hitler, with perhaps a touch of Al Capone for seasoning. In all this, he is seen as a true representative of the expansion of western Europe, which from the fifteenth century onward brought almost the whole world into the orbit of European civilization and made it subject to the rule, or at least the influence, of European powers.

Those who are familiar with modern thought will know

that in this expansion of western Europe there was a special quality of moral delinquency, absent from such earlier, relatively innocent expansions as those of the Mongols, the Huns, the Ottomans, the Arabs, the Aztecs—even from the concurrent expansion of the Muscovites, whose manifest destiny also took them from sea to shining sea, that is, from the Baltic to the Pacific. In this connection, it may be useful to recall that the expansion of Europe in the modern centuries took place at both ends: the competing western nations across the seas to America, Africa, and South and Southeast Asia; the Russians, with helpers not rivals, across the steppes to the Black Sea, the Caspian, Central Asia, and the borders of China.

In some places, the advancing Europeans came to what they regarded and treated as empty lands: more precisely, lands sparsely inhabited by largely nomadic populations. In such lands, the Europeans were able, by conquest, colonization, and settlement, to create the English- and French-speaking societies of North America and the Russian society of North Asia, otherwise known as Siberia. The French in North Africa and the Dutch and British in South Africa attempted but did not achieve the same result.

In other places, the Europeans encountered and clashed with existing civilizations. Five of these were of major extent and importance, two in the Western Hemisphere and three in the Eastern Hemisphere. I refer, of course, to the three great civilizations of China, India, and Islam. The three of them, with their own zones of expansion and influence, covered almost the whole of Asia and a large part of Africa. In due course, all three were themselves subjected to the devastating impact of an expanding and conquering Europe.

In the discovery of America, as in those other parallel processes by which European explorers, merchants, missionaries, and others "discovered" Asia and Africa, there was always an important element of conquest. The Spaniards and others who followed in Columbus's wake were accurately designated by their contemporaries as the *conquistadores,* the conquerors of a new world. This conquest neither started nor ended with the discovery of America, but was part of a much larger process, beginning centuries before Columbus sailed and continuing for centuries after his achievement. Indeed, the self-same year, 1492, was inaugurated with such a conquest, the conquest of Granada. On January 2 of that year, the combined armies of the Catholic monarchs Ferdinand of Aragon and Isabella of Castile entered the previously surrendered city of Granada, the last stronghold of Muslim power in Spain, and thus achieved the final Christian victory in the eight-century-long struggle between Christianity and Islam for the mastery of the Iberian Peninsula. It was the first of the three major events of what was to be, for Spanish Christians, an *annus mirabilis*—a prerequisite without which neither the second nor the third event would have been feasible.

The encounter between Europe and Islam differed profoundly on both sides from the parallel encounters with India and China. Fifteenth-century Europeans knew very little about these two countries. There were a few moldering remnants of knowledge left over from Greek and Roman times, occasional reports brought back by travelers, priests, and merchants who, driven by the two compelling motivations of faith and greed—perhaps the strongest motivations in human history, especially when combined—had ventured into these remote lands and brought back strange tales of what they had seen or heard. But they knew very little. China and India were hardly more than names linked to certain commodities that came from those remote places. There was no real knowledge, no past record of dealings, and, in consequence, no set attitudes. If Europeans knew little about India and China, Indians and Chinese knew even less about Europe, for even fewer travelers ventured westward and even less knowledge was brought back east. And of course, between Europe and the peoples of the Americas, there was total, perhaps blissful, mutual ignorance.

The history of relations, and therefore of perceptions and attitudes, between Europe and Islam was very different. Europeans and Muslims knew a great deal about each other; some of it was even accurate. There is also an important, indeed a crucial, difference between Islam and the two other Old World civilizations. China is a place; India is a place; Islam is a religion. The civilizations of China and East Asia, of India and Southeast Asia, and, here we may add, of pre-Columbian America may have been very advanced and very sophisticated, but they were and remained essentially regional. Apart from a natural processes of expansion and influence in immediately adjoining areas, they were limited to one place and, to a remarkable extent, to one culture and even to one race.

In contrast with these ethnically defined and geographically limited civilizations, Islam was in principle universal in its beliefs, its self-perception, and its ambitions. The Muslim believed himself to be the fortunate possessor of God's final revelation and saw it as his duty and his privilege to bring God's truth to the rest of mankind, if necessary, by *jihād,* by war "in the path of God." In this, Islam differed sharply from the other civilizations of Asia, Africa, and the Americas but resembled Christianity, which, from its beginnings, was inspired by its own sense of privilege and mission. Until the end of the Middle Ages, the Muslims had gone considerably further than the Christians in giving practical effect to their aspirations. Despite its universalist aims and claims, Christianity had remained basically European, and even its holy wars, its Crusades, were waged to defend or recover Christian territory. The one exception outside Europe, the Christian kingdom of Ethiopia, was remote and little known and had almost no impact on Christian self-perception. Much the same is true of the Christian minorities under Muslim rule in Southwest Asia. For most practical purposes, Christianity was as European as Hinduism was Indian—the religion of one part of the world, relatively speaking quite a small one, where all the inhabitants were of one race and culture and belonged to a limited number of mostly interrelated ethnic groups.

Islamic civilization, in contrast, was the first that can be called universal, in the sense that it comprised people of many different races and cultures, on three different continents. It was European, having flourished for a long time in Spain and southern Italy, on the Russian steppes, and in the Balkan Peninsula. It was self-evidently Asian and also African. It included people who were white, black, brown, and yellow. Territorially, it extended from southern Europe into the heart of Africa and into Asia as far as and eventually beyond the frontiers of India and China. Since the mission of the Muḥammad in seventh-century Arabia and the expansion of his followers into the Mediterranean world, Islam was the neighbor, the rival, and often the enemy of Christendom.

There had been an earlier missionary religion with universalist aspirations that had preceded both Islam and Christianity. But Buddhism had long since lost its missionary ardor and had become a regional religion in East and Southeast Asia.

During the centuries that European historians called medieval, each of the two surviving missionary religions, Christianity and Islam, rightly saw the other as its principal rival. I referred earlier to the two compelling motives of faith and greed, which sent missionaries and traders all over the world. In medieval Christian Europe, in confrontation with Islam, there was a third motive, perhaps more compelling than either of the other two, and that was fear.

For roughly a thousand years, from the first irruption of the Muslim armies into the Christian lands of the Levant in the early seventh century until the second and final Turkish withdrawal from the walls of Vienna in 1683, European Christendom lived under the constant and imminent menace of Islam. The very first Islamic expansion took place largely at Christian expense: Syria, Palestine, Egypt, and North Africa were all Christian countries, provinces of the Christian Roman Empire or subject to other Christian rulers, until they were incorporated in the realm of the caliphs. Their lands were lost to the conquering armies of Islam, their peoples to the militant faith of the conquerors. The Muslim advance continued into Europe not once, but three times. The first wave of Muslim expansion into Europe began in the early years of the eighth century and for a while engulfed Spain, Portugal, southern Italy, and even parts of France. It did not end until 1492, with the defeat and extinction of the last Muslim state on west European soil. The second wave struck eastern Europe when the Mongols of the Golden Horde, who had established their domination over Russia and most of eastern Europe, were converted to Islam and subjected Muscovy and the other Russian principalities to the suzerainty of a Muslim overlord. That, too, was ended after a long and bitter struggle by a Christian reconquest and the withdrawal of the Islamized Tatars from Russia. The third wave was that of the Seljuk and Ottoman Turks, who, after conquering Anatolia from the Byzantine Empire, crossed into Europe and established a mighty empire in the Balkan Peninsula. In the course of their advance, the Turks captured Constantinople and twice laid siege to Vienna, while the vessels of the Barbary corsairs carried the naval *jihād* as far as the British Isles and, on one occasion, even Iceland. It was the second siege of Vienna and the defeat and retreat of the Ottoman forces that marked the real turning point in the relations between the two religions and civilizations.

The thousand-year-long Muslim threat to Europe was twofold, military and religious, the threat of conquest and of conversion. West of Iran and the Arabian Peninsula, the vast majority of the earliest converts to Islam in the Levant and North Africa were converts from Christianity. This process of conversion continued in the Muslim-ruled lands of Sicily and Spain. Great efforts were made by medieval Christian European scholars to understand the rival religion, in order to refute it and thus protect their own flocks. Thanks in large measure to the tolerance that the Christian communities and their leaders enjoyed under Muslim rule, they were in the main successful in this endeavor, though significant numbers did indeed adopt the faith of their conquerors.

There is a pervasive sense of fear in discussions, by Europeans, of Islam and of the Muslim peoples whom they encountered: Moors and Saracens, Tatars and Turks. In poetry and polemic, in history and literature, they reflect the consciousness of a Christian Europe besieged and threatened by a mighty and expanding Islamic world, a Europe that in a sense was defined and delimited by the frontiers of Muslim power in the east, the southeast, and the south. The medieval Muslim perspective of Europe, as reflected in literature, is very different. Judging from the rare and rather disdainful references in Muslim writings, western Europe must have appeared to them rather as Central Asia or Africa appeared to Victorian Englishmen.[1] For Muslims, the land beyond the northwestern frontier of Islam was a remote and unexplored wilderness inhabited by exotic and picturesque tribes with dirty and nasty habits, possessing a very low level of culture and professing a superseded religion, and with few commodities of any value to offer, apart from their own people, who might be brought to some minimum level of civilization through the divinely ordained institution of slavery.

There were a few intrepid explorers who ventured into darkest Europe and left accounts of their travels, but their reports seem to have had little impact. Muslims were aware of Byzantine Europe: They knew and respected the civilization of the ancient Greeks and, to a much lesser extent, of the Christian Greeks of Byzantium, but they had no respect, nor indeed was there any reason why they should have had any respect, for central and western Europe, which in medieval times was at an incomparably lower level of civilization, moral and material, than the Islamic lands. Nevertheless, there was an awareness in the Islamic world that these were not merely barbarians, like some other neighbors of the Islamic world in Asia and Africa. They were followers of a rival religion, with a rival political system and a rival claim to bringing a universal message and imposing a universal law on all humankind.

The world was divided into the House of Islam, where the Muslim faith and law prevailed, and the House of War, where they did not, and between the two there would be a perpetual state of war, interrupted only by truces, until the Word of God was brought to all humanity. For most Muslim writers, Christendom—first Byzantine and then European—was the House of War par excellence.

Between Islam and Christendom there was inevitably great and continuing hostility, but it was not due, in accordance with currently fashionable notions, to misperception and misunderstanding. On the contrary, the two understood each other very well, far better than either of them, in their other encounters, could understand the more remote civilizations of Asia and, later, pre-Columbian America. As well as a shared or, rather, disputed mission and domain, Islam and Christendom had a great shared inheritance, which drew on common sources: the science and philosophy of Greece, the law and government of Rome, the ethical monotheism of Judaea, and, beyond all of them, the deeply rooted cultures of the ancient Middle East. Christians and Muslims around the Mediterranean could find a common language in both the figurative and the literal senses. They could communicate, they could argue, if only to disagree; they could translate, as they did, both ways. All this would have been difficult, if not impossible, between Christians or Muslims, on the one hand, and exponents of the civilizations of India or China, on the other.

True, they denounced each other as infidels, but in so doing, they revealed their essential similarity, even kinship. Both claimed possession of universal and final truths, of God's last word, which it was their duty to bring to the rest of the world. For the accomplishment of this divinely ordained task, they expected permanent and final reward in paradise and might in the meantime be accorded some provisional material recompense in this world. Those who differed from them would similarly suffer, both in this world and in the next. Christian and Muslim ideas about judgment, retribution, and reward in an afterlife were basically similar, though not identical. Their heavens were significantly different, but their hells were much the same. All this would have been meaningless to a Hindu, a Buddhist, or a Confucian.

In virtually all modern historical narratives, there is an accepted periodization, a sequence from ancient through medieval to modern. "Ancient" begins with the history of the lands and peoples mentioned in the Bible, reaches its peak in the world of Greece and Rome, and ends with the decline and fall of the Roman Empire in the west. "Modern" denotes the period beginning in the late fifteenth century, when a resurgent and expanding Europe inaugurated a new era and a new civilization that eventually embraced the whole world. Between these two vast and flourishing civilizations, medieval Europe served like the connecting bar of a dumbbell.

But medieval Europe was not the only route connecting the ancient with the modern world. Indeed, Islamic civilization in its prime might well have seemed a more promising transition from a shared antiquity to a divided modernity. A comparison between medieval Christendom and medieval Islam would surely have shown the Islamic world as offering the high road from ancient to modern civilization. The Islamic world was rich and vast, covering an immense area with a great variety of peoples and a great wealth of resources. Like Western Christendom, it enjoyed the heritage of Hellenistic civilization, but with a far greater mastery of its science and philosophy,[2] and had greatly enriched this heritage through its contacts with other civilizations and through its own creative endeavors. To name but two examples, the importation of paper and then of papermaking from China and the acceptance of positional numbering and the zero from India prepared the way for a great scientific and literary renaissance, centuries before the European movement for which that name is customarily used. Christian Europe, in contrast, was poor in resources, limited and local in its outlook, and in most, though not all, respects, far below the levels of Islamic achievement.

There were other differences: Medieval Europe was not only small in extent, but also narrow in outlook, with a remarkable lack of tolerance, not only of other religions but even of divergent forms of its own. The Islamic world, in contrast, was diverse in composition and pluralistic in character. Muslims were willing to tolerate significant differences in practice and even belief among themselves; they were also willing to concede a certain place in society to other, approved religions.[3] This has sometimes been misrepresented in modern times as equality. It was not, of course, equality, and the very idea of such equality in a medieval society is absurdly anachronistic. The granting of equal rights by believers to unbelievers would have been seen, on both sides of the Mediterranean, not as a merit but as a dereliction of duty.

Islamic society did, however, grant toleration; there was a willingness to coexist with people of other religions who, in return for the acceptance of a few restrictions and disabilities, could enjoy the free exercise of their religions and the free conduct of their own affairs. There is no true equivalent to this tolerance in Christendom until the Wars of Religion finally convinced Christians that it was time to live and let live. During the eight centuries that Muslims ruled part of the Iberian Peninsula, Christians and also Jews remained and even flourished. The consequences of the Christian reconquest, for Jews and Muslims alike, are well known.

Only in one respect did Europe show an obviously greater diversity—in language. All over southwestern Asia and northern Africa, one language, Arabic, met the needs of religion and law, commerce and culture, government and science. While the vernaculars varied enormously from country to country, a standard written Arabic served as the medium of both practical and scientific communication throughout the Arab and, to a lesser extent, the whole Muslim world. As far away as Iran, the great fourteenth-century Persian historian Rashīd al-Dīn commented in astonishment that the Franks spoke twenty-five different languages and that no people among them understood the language of any other.[4]

And yet despite all this, it was the Islamic route, not the Western route, that was "the road not taken." By the late fifteenth century, certain basic changes were already taking place in both the Christian and Muslim worlds, which eventually ensured the rise and triumph of the one and the decline and defeat of the other. The struggle between them was long and bitter; centuries passed before the outcome was clear and recognized on both sides. The rise of the West has been much studied, but the waning of Islamic power has received little serious scholarly attention, and at this stage one can do little more than list the phenomena associated with these changes, with only the most tentative evaluation of their places as causes, effects, or merely symptoms.

It was, of course, in warfare, and more particularly in weaponry, that the growing disparity between the two sides first became apparent. In the early days of Islam, Muslim armies carried all before them, bursting across Asia to India and China, across Africa to Morocco and far into Europe. The Muslims' further advance was halted not so much by the resistance of their enemies as by the growing strain on their own system and especially on its far-flung communications. Their weaponry was as good as that of their opponents and sometimes even innovative, as, for example, in the use of petroleum, plentiful in the Middle East, to make explosive mixtures. These, in suitably designed pots, served as projectiles in both naval and siege warfare. Their body armor and siegecraft were perhaps not as good as those of the Christians, but the Muslims were quick to learn.

The notion of war for the faith that inspired the Muslim conquests was not entirely unknown to Byzantine Christendom. The emperor Heraclius (610–641) presented his struggle against the Persians as a Christian holy war, and Nicephorus Phocas (963–969), who reconquered northern Syria from the Muslims, is quoted as recognizing the power of the Muslim *jihād* and desiring a Christian equivalent. More often, Byzantine writers speak of such war with incomprehension, even distaste, and see its exponents as driven more by the appetite for booty than by religious devotion. The idea of a Christian holy war evoked a better, albeit delayed, response in Western Christendom. After an Arab raid on Rome and Ostia in 846, a synod was convened in France, and an appeal was issued to the kings of Christendom to form a joint Christian army, under French leadership, to make war against the enemies of Christ. Pope Leo IV, perhaps influenced by Muslim ideas, promised a heavenly reward to all who died in such a war. Some decades later, Pope John VIII (872–882), in answer to a request from his bishops, promised remission of sins to those who fought in defense of the Holy Church of God and for the Christian state and faith and who died "contra paganos atque infideles strenue dimicantes" (strenuously struggling against pagans and infidels).[5]

Even in the West, the response to the calls for a holy war in defense of Christendom was slow, and for some time the results were inconclusive. The decisive change began in the second half of the eleventh century. In 1061, the Normans captured Messina and proceeded to reconquer Sicily for Christendom. In 1085, with the capture of Toledo by Alfonso VI, the Christian reconquest of the Iberian Peninsula was well under way. After some setbacks, it was resumed, and the great victory of Las Navas de Tolosa in 1212 broke the back of Muslim power in the peninsula. Only a few local Muslim states remained, which the advancing Christians overwhelmed one by one. Granada was the last. In the parallel reconquest of eastern Europe, Dmitri Donskoi, prince of Moscow, won a great victory at the Battle of Kulikovo in 1380 against the Tatars. As in Spain, a long, hard struggle followed, but in 1480, Ivan III, known as the Great, defeated the last Tatar advance on Moscow and finally threw off "the Tatar yoke."

There were limits to the successes of the *reconquista*. The

Crusades, often nowadays interpreted as an early Western venture in aggressive imperialism, can be more accurately seen in the context of the time as part of the *reconquista,* inspired by the same motives and directed against the same enemy. Christianity, after all, was much older and much more deeply rooted in the Middle East than in southern Europe. The crusade, a long-delayed Christian response to the *jihād,* was an attempt to recover by holy war what had been lost by holy war, and what could be more important for Christians than the Christian Holy Land, lost by the Byzantine emperors to the Muslim caliphs in the seventh century? The *reconquista* succeeded in Iberia and Russia; it failed in the Levant, where it was overwhelmed by the new and rising power of the Turks. These new champions of Islam, who had already engulfed the once Greek and Christian land of Anatolia, would soon launch the third great attack on Europe.

Despite these and other Turkish successes, the disparity in favor of Christendom grew ever wider. Christian victories were due in no small part to their superior weaponry and technology and to the stronger economies that supported them. Gunpowder was invented in the Far East, where it seems to have been used principally for firework displays. Bypassing the Middle East, it was brought to Europe, where it was adapted to a new and deadly purpose—firearms. These gave an immense and often decisive advantage to Europeans in their warfare with others, most obviously in the New World, but also to a growing extent in their encounters with the civilizations of the Old World and even with the empires of Islam.

Some Muslim states either rejected firearms or made very

little use of them.[6] The once mighty Mamluk sultanate of Egypt, for example, regarded these unchivalrous weapons with contempt and confined their use to small units manned by socially inferior elements. Not surprisingly, the Mamluks were unable to resist either the Portuguese coming from the south or the Ottomans coming from the north. Their attitude is well expressed by the Egyptian historian Ibn Zunbul, writing after the Ottoman conquest of Egypt. According to him, a captured Mamluk emir confronted the Ottoman sultan Selim and told him:

> You have patched up an army from all parts of the world: Christians, Greeks, and others, and you have brought with you this contrivance, artfully devised by the Christians of Europe when they were incapable of meeting the Muslim armies on the battlefield. It is this musket, which even if a woman were to fire it, she could hold up so many men. Had we chosen to use this weapon, you would not have preceded us in its use. But we are people who do not discard the *sunna* of our Prophet Muḥammad, which is the *jihād* in the path of God, with sword and lance. . . . Woe to you: how can you shoot with firearms against those who testify to the unity of God and the Prophethood of Muḥammad?[7]

Only the Ottomans among Muslim states made full and effective use of musketry and artillery, but even they, as Ibn Zunbul had noted, were dependent on Western technology and, to an increasing extent, relied on Western renegades and mercenaries to equip and direct their artillery.

The West also enjoyed a growing superiority at sea. West European ships were built for the Atlantic, and they were therefore bigger and stronger than those of the Muslims, built for the Mediterranean, the Red Sea, and the Indian Ocean. The European ships were more maneuverable, could travel greater distances, and, above all, could carry a much heavier armament. A Spanish galleon or a Portuguese *carrack* could easily outmatch any vessel owned by the Muslim powers, whether for transporting cargo or for conducting naval warfare. And they, in turn, were outpaced and outmaneuvered by vessels built for the rougher seas of the north. An Ottoman historian of the time, Selaniki Mustafa, gives a vivid picture of the impression made by the arrival of an English ship, bringing a new ambassador, in 1593:

> A ship as strange as this has never entered the port of Istanbul. It crossed 3700 miles of sea, and carried 83 guns, besides other weapons. The outward form of the firearms was in the shape of a pig. It was a wonder of the age, the like of which has not been seen or recorded.[8]

The Muslim world, too, had its Atlantic coastline, and one may wonder why the Moroccans did not sail the high seas or discover America. One obvious answer is the absence, on that coastline, of the ports and estuaries that facilitated—even necessitated—the development of European seafaring and seamanship. Another answer may be that the Moroccans had their seas pretty much to themselves, in this differing from the Spaniards and the Portuguese, the French, the Dutch, and the English, who honed their naval skills in incessant warfare against one another. In this, as in some other respects, European disunity was to prove a long-term advantage.

The greater technological sophistication of western Europe was not limited to weaponry, but extended to economic production. A good example is the use of the mill, at first and for long the only device for generating energy other than human or animal muscle. Mills, even primitive windmills and watermills, are large and visible and fixed. They cannot easily be hidden or disguised or moved. They were therefore the delight of the tax collector and remain the delight of the modern historian who relies on tax records. A comparison between the Domesday Book and the Ottoman imperial registers made by my Princeton colleague Charles Issawi revealed the astonishing fact that there were proportionately more mills in Norman England than in the central Ottoman lands in the days of Süleyman the Magnificent.[9]

Another technological innovation of immense importance was printing. This too, like gunpowder, was of East Asian origin, and this too, apart from a brief and disastrous experiment with paper currency by a Mongol ruler in Persia,[10] bypassed the Middle East. Spanish Jewish refugees who arrived in Turkey in 1492 brought printing presses with them and were granted permission by the sultan to print books in the capital and other cities, on condition that they not print in Arabic characters. Similar permission was later given to the Christian minorities. Until the eighteenth century, books were printed in the Ottoman lands in Hebrew, Greek, Armenian, Syriac, and occasionally Latin characters, but not in the script used by the Turks and all their Muslim subjects.[11] This double and decisive rejection of printing, both when it came from the East and when it was reintroduced from the West, is all the more remarkable when compared with the eager and effective acceptance of paper some centuries earlier.

Gunnery and printing are but two of the more dramatic aspects of this mounting disparity. Their rapid development and use in the West, as contrasted with their tardy, reluctant, and, for long, ineffectual adoption in the Islamic lands, must surely be due in part at least to deeper social and cultural causes, the study of which has barely begun.

One such cause has already been indicated: the inability of the slave military society, of the kind that had become normal in Muslim states in the later Middle Ages, to adapt to change. Intimately associated with the slave military society was a social order requiring the segregation and subordination of women. Let me quote a more recent authority. In a speech delivered in 1923, the founder of the Turkish Republic, Kemal Atatürk, said: "If a society contents itself with modernizing only one of the sexes, that society will be weakened by more than one half."[12] The women of Christian Europe were very far from achieving any kind of equality, but they were not subject to polygamy or legal concubinage. Even the limited measure of freedom and participation that they enjoyed never failed to shock a succession of Muslim visitors— all of them male—to Western lands.[13] Western civilization was richer for women's presence; Muslim civilization, poorer by their absence.

A few other anniversaries may suggest some different aspects of the disparity between the two societies. The year 1492 saw the death of Lorenzo de' Medici, that archetypal Renaissance prince, who played a major role in the patronage of the new culture and learning and in the conduct of new-style business, and—a point of some relevance—who pioneered the new Western craft of foreign policy, which was to play so large a part in the subsequent expansion of the European powers.

The same year saw the publication, by the Spanish humanist Antonio de Nebrija, of his *Grammática castellana,* the first attempt to stabilize and standardize a European vernacular and turn it into a literary language. Until then, the situation in Latin Europe and in the Arab world had been roughly similar, both using an artificial language—the one, medieval Latin; the other, postclassical Arabic—for both official and literary writing, while speaking in their respective vernaculars. The French historian Bernard Vincent has argued that the establishment of the Castilian dialect as standard Spanish contributed greatly to the rise and efflorescence of Hispano-American civilization,[14] and he cites the remarkable words of an Aragonese scholar, Gonzalo García de Santa María:

> Since the royal power is today Castilian, and since the excellent king and queen who rule us have chosen to make the realm of Castile the base and seat of their states, I have decided to write this book in Castilian, because language, more than all else, accompanies power.[15]

Others have argued, in a sense along similar lines, that the rejection of the vernaculars and the continued use of a form of classical Arabic in all written material has been an obstacle to cultural progress in the Arab lands.

Islam had had its own Renaissance, in the tenth and eleventh centuries, with the revival of ancient learning and the creation and expansion of new learning. The European Renaissance had virtually no effect in the Islamic lands, where, in the words of the Turkish historian Adnan Adıvar, "the scientific current broke against the dikes of theology and jurisprudence."[16] The Islamic world in the fifteenth and sixteenth centuries had no Renaissance, neither did it have a Reformation.

But it remained immensely powerful and still confronted Europe at both ends of the Mediterranean. In 1453, the Turkish conquest of Constantinople dealt a devastating blow to Eastern Christendom and posed a major challenge to the West. By 1478, the Turks ruled the lands of the Greeks, the Albanians, the Romanians, and the south Slavs, and their advance forces had reached the outskirts of Venice. In 1480, with the capture of Otranto, they gained a foothold in Italy. The Spanish conquest of Granada was a major victory, which proved decisive in saving southwestern Europe. It was not, however, decisive in the larger confrontation between Christendom and Islam. In southeastern Europe, the Muslim threat remained and even grew; it did not end until two centuries later, with the final retreat of the Turks from Vienna.

In the meantime, the Catholic monarchs of Spain, soon after their victory over the Muslims, turned their attention to another enemy: the Jews. In politics and war, the Jews were insignificant. In business and culture, they could even be useful. But in religion, they were seen as offering the most intimate and most sensitive of challenges.

CHAPTER 2
Expulsion

O N 13 APRIL 1602, a concerned citizen of Venice petitioned the republic against the proposal to establish the Fondaco dei Turchi on a regular basis. This institution, which provided board and lodging for visiting Turkish merchants, was, according to the petitioner, objectionable and dangerous for a number of reasons. The presence of large numbers of Turks in one place would inevitably lead to the building of a mosque and to the worship (the petitioner evidently knew little of the Islamic religion) of Muḥammad. This would be an even greater scandal than that already caused by the presence of Jews and Protestant Germans. For one thing, the dissolute behavior of the Turks would have a demoralizing effect. For another, rather more important, the Fondaco would facilitate the political designs of the Turks, who, "guided by a single sultan and equipped with great naval power," could thus do far more harm to Venice than the Jews, who were "without any chief or prince" and who were "downtrodden by the world."[1]

The linking of Muslims and heretics was not uncommon. A story, widespread in the Western Middle Ages, even presents Muḥammad himself as a renegade cardinal who, out of pique at having been passed over in a papal election, went off to Arabia and started a rival religion.[2] Even apart from such absurd fantasies, Catholic polemics against the Protestants sometimes present Islam as an early Reformation and allude pointedly to such common features as hatred of images, rejection of the Trinity and of Mariolatry, and, on a different level, alleged collusion between the sultan and the Protestant princes. But the linking of Muslims and Jews as enemies of the church and the faith is much older and much more widespread and goes back at least to the time of the First Crusade.[3]

In the so-called Dark Ages, Jews in Christian Europe had enjoyed a fair measure of tolerance. Their troubles began when the Crusaders set forth to do battle with the infidel in the Holy Land and thought it appropriate to begin with the infidels at home. In the words of a contemporary Hebrew chronicler:

> Now it came to pass, that as they passed through the towns where Jews dwelled, they said to one another: "Look now, we are going a long way to seek out the profane shrine, and to avenge ourselves on the Ishmaelites, when here in our very midst are the Jews, they whose forefathers murdered and crucified Him for no reason. Let us first avenge ourselves on them, and extermi-

nate them from among the nations, so that the name of Israel will no longer be remembered—or let them adopt our faith."[4]

From the late eleventh century onward, many Jewish communities in central and western Europe were offered the choice of conversion or death, or, as they would have seen it, of apostasy or martyrdom. Often, these outbreaks were instigated by the very persons or authorities who were calling the faithful to a new crusade. They include such celebrated figures as the fifteenth-century Franciscan preacher and inquisitor St. John of Capistrano, remarkably successful as a persecutor of heretics and Jews, rather less so as an instigator of crusades. He was later canonized, and his name still graces the map of California.

In those days of heightened religious awareness, when Christendom seemed to be endangered both at home and abroad by the triple threat of schism, heresy, and unbelief, the tolerance previously accorded to divergent opinions, always precarious, dwindled to a vanishing point. Particularly in the lands newly recovered for Christendom, Jews and Muslims were seen as a continuing threat to hard-won Christian unity and independence. In western Europe, Jews and Moors and, in eastern Europe, Jews and Turks were commonly named together in polemic, in exhortation, and even in regulations, local, royal, and papal, as the enemies of Christendom. Even their daily habits appeared to confirm this perception. As late as 1670, a British naval chaplain aboard a ship on the Mediterranean, who whiled away his time by keeping a diary and writing bad verse, illustrated one aspect of this association:

God save King Charles, the Duke of York,
The Royal Family;
From Turk and Jew who eat no pork,
Good Lord deliver me.

In this atmosphere, it is not surprising that perfervid imaginations sometimes discerned patterns of conspiracy: occasionally, the Muslim as the agent of the Jew; much more frequently, the Jew as the agent, emissary, or at least sympathizer of the Muslim. As late as 1877 to 1888, Liberal opponents of Disraeli's pro-Turkish policy resorted to this argument. According to a Liberal member of Parliament and writer, T. P. O'Connor: "His [Disraeli's] general view . . . upon the question of Turkey is that as a Jew he is a kinsman of the Turk, and that as a Jew he feels bound to make common cause with the Turk." An eminent historian, E. A. Freeman of Oxford, went even further: "Throughout Europe, the most fiercely Turkish part of the Press is largely in Jewish hands. It may be assumed everywhere, with the smallest class of exceptions, that the Jew is the friend of the Turk." Even Gladstone, the leader of the Liberal Party, in a conversation with the Duke of Argyll, remarked, "I have a strong suspicion that Dizzy's crypto-Judaism has had to do with these policies."[5] If even Victorian English Liberals could think and speak along these lines, small wonder that others, in less tolerant times and places, should harbor even darker suspicions, and be ready to act on them.

In January 1492, after their victory at Granada, the Catholic monarchs of newly united Spain saw themselves as facing two dangers. One was political and military—the danger of a Muslim counterattack. At the other end of the Mediterranean, a new and mighty Muslim empire, triumphant after its capture of Constantinople some forty years previously, was extending both its military and its naval power westward. Only a few miles away, in North Africa, there were still formidable Muslim forces, as was demonstrated in 1578, when the Moroccans inflicted a crushing defeat on the Portuguese at the Battle of the Three Kings. Defensive and perhaps preemptive measures were clearly necessary against a possible Muslim return.

The other danger was religious. In this respect, Islam was no longer seen as a dangerous enemy. In a theological calculation, it could, so to speak, be discounted. Being subsequent to the Christian dispensation, it was in Christian terms necessarily false and could be dismissed as a heresy or an aberration. It could even be incorporated in a Christian eschatology as the "beast" of Revelation. The attraction of Islam to converts, so powerful in the days of Muslim advance, waned with the Muslim retreat. Even the mighty victories of the Ottomans did not, as had the earlier victories of the caliphs, persuade large numbers of the conquered that Islam was indeed God's true religion.

The Jews, however, were a different matter. Politically and militarily, they were unimportant: As the Venetian petitioner had correctly noted, they were a leaderless and downtrodden people, incapable even if they wished it of inflicting any harm. The tacit religious challenge that Judaism offered to Christianity was another matter, however. Being pre-Christian and not post-Christian, it could not be dismissed as a heresy or an aberration. The Hebrew Bible, renamed the Old Testament, had been adopted by the Christians, who added a new testament to it, explaining how Christ had come to complete the revelation and to fulfill the promises that God had given to the Jews. By this logic, the Jews should have been the first to welcome and to accept the new dispensation and to merge their identity in the Christian church as the new beneficiary of God's choice.

Many Jews did indeed accept this view and were the nucleus of what later became Christendom. But the persistent refusal of others and the survival of Judaism as a separate religion were seen by many Christians as impugning the central tenets of their faith. Jews, unlike Muslims, could not be accused of not knowing the Old Testament or of being unaware of the Choice and the Promise. Their unwillingness to accept the Christian interpretation of these books and of these doctrines thus challenged Christianity in a most sensitive area.

The challenge and the implied insult were even more wounding when large numbers of Jews, forcibly converted to Christianity and instructed in their new faith, nevertheless persisted in preserving their old faith in secret and even in transmitting it to their children and grandchildren, baptized at birth and educated as Christians. One of the arguments for the expulsion was that the continued presence of contumacious, unconverted Jews impeded the true conversion of those who had accepted baptism under compulsion, and created an ever present danger of apostasy.

There was another and, in the long run, more insidious threat that the Jews posed, at least to ecclesiastical authority. It was correctly identified by St. John of Capistrano, who, in some of his sermons delivered in Germany and in Vienna, told of a "deceitful" notion that Jews were trying to spread among Christians: "The Jews say that everyone can be saved in his own faith, which is impossible."[6] For once, St. John of Capistrano did not malign the Jews or their faith. According to rabbinic teaching, before the Ten Commandments were revealed to Moses, there were seven commandments revealed to Noah and binding on all humankind.[7] The first two might be called religious; they prescribe belief in a single God and forbid polytheism and graven images. The others, prohibiting murder, robbery, sexual offenses, cruelty to animals, and the like, form what one might call a code of natural law, which defines true righteousness. According to a well-known rabbinic dictum, the righteous of all peoples have a share in paradise. For St. John of Capistrano and others like him, it was easy to understand the triumphalism of Islam, so akin to their own, and to confront it appropriately—that is, on the battlefield. The religious relativism of their unarmed Jewish neighbors posed a different challenge, requiring a different response.

After their victory at Granada and the union of the Spanish kingdoms, the Catholic monarchs had achieved political and military unity. They had not achieved religious unity, since large numbers of Jews and still larger numbers of Muslims remained in the lands under their rule. No doubt for good practical as well as ideological reasons, they turned their attention first to the Jews, the most threatening to Christian teachings, and the most vulnerable to Christian power.

The expulsion of the Jews—"this leaderless and downtrodden people"—was relatively safe and easy. It also provided useful experience for the later, greater, more difficult, and vastly more important task of expelling the Muslims.

The Edict of Expulsion of the Jews was signed and proclaimed in Granada on 31 March 1492 and promulgated on 29 April. All Jews were to accept baptism or leave the kingdom by 31 July.

The expulsion of the Jews from Spain in 1492 by decree of Ferdinand and Isabella, though by far the best known such expulsion, was by no means the first. In the late thirteenth and early fourteenth centuries, Jews were expelled by royal decree from the kingdoms of Naples, England, and France, as well as from many cities and principalities. Nor was it the last. Only a few years later, in 1496, the Jews of Portugal, plus an estimated 100,000 Spanish Jews who had found a brief and illusory refuge among them, were subjected to an even harsher expulsion.

Faced with such agonizing choices, great numbers of Spanish and Portuguese Jews found an uneasy and dangerous solution to their problem, by combining the public acceptance and private rejection of the faith that was thus imposed on them. The converts were officially known as *conversos* or *nuevos cristianos*—"new Christians"; unofficially, they were often described by the *viejos cristianos*—old Christians—as *marranos*. The Spanish word *marrano*—literally "swine" and figuratively "a person of swinish character or habits"—came to be applied more particularly to those new Christians who were suspected of practicing Judaism in secret. Marranism in this sense—the secret adherence to one faith, combined with the public avowal of another—is by no means limited to the Iberian Peninsula and has many parallels in other times and places. But nowhere, with one significant exception, did it achieve the scale and importance of the crypto-Jews of Spain and Portugal.[8]

The first anniversary of 1492, the capture of Granada, is

still celebrated in good faith as a great victory for the Spanish nation and the Christian church. The second quincentennial of 1492, the expulsion of the Jews, was not celebrated,[9] at least not in public, but was commemorated, in a becoming spirit. In at least one place, however, in Istanbul, there was not just recollection in 1992, but celebration: not, indeed, of the departure of the Jews from Spain, but of their arrival in Turkey, where by far the largest group of Spanish and Portuguese exiles found permanent refuge and new homes. The year 1992 saw many commemorative events, organized by the descendants of those who had arrived five hundred years previously, with the assistance and cooperation of the descendants of those who had accepted them.

The Jews who fled from Spain and Portugal made their way to many countries: to France, Holland, and later England; to Italy, where ironically a number of these refugees from Catholic persecution in Spain and Portugal found refuge in the states of the church. In Venice, where reference to *marrano* origins might have endangered their freedom or even their lives, they were known as Ponentini, westerners, as opposed to the Levantini, who came from the east. In the years that followed the expulsion, some even crossed the Atlantic and found new homes and new lives in the European colonies in the Americas. There were notable figures among the western refugees and their descendants. But by far the most important group were those who went to the Islamic lands and, more particularly, to the vast realms of the Ottoman Empire, in Europe, in Asia, and later also in Africa.[10]

The Jews had a special reason for preferring an Islamic country. Many of them had stayed at home in Spain and Portugal and accepted baptism, waiting for a suitable oppor-

tunity at a later date to emigrate and revert to Judaism. In any
Christian country, even the most tolerant, such an apostasy
was punishable by death. In Muslim lands, too, apostasy
from Islam was a capital offense, but apostasy from Christianity to Judaism was a matter of indifference. Muslim jurists
in any case often took the view that a forced conversion was
not valid and that renouncing such a conversion did not
therefore constitute apostasy.[11] The Jewish *marranos* thus
had an obvious interest in preferring a Muslim to a Christian
environment for their reversion to their old faith. The difference was grimly exemplified in 1555/1556, when a number
of Portuguese Jews, living openly as such in Ancona, in the
Papal States, were accused of being *marranos* and charged
with apostasy. Under questioning, some confessed, abjured
their Judaism, and were rewarded by having their punish-
ment commuted to service for life in the galleys. The rest—
twenty-four men and one woman—were consigned to the
secular arm and then strangled and burned. Even a strong
diplomatic intervention by the Ottoman sultan failed to se-
cure their release.[12]

The Jews who settled in the Ottoman lands were by far
the largest group of those who had fled from Spain and Portugal. They were also the largest single group among the
many refugees from Europe who fled to Turkey, including
Jews from other European countries, as well as Christians
fleeing from persecution as schismatics or heretics. In the case
of the Jews, there is, in addition, serious evidence that their
arrival was not merely permitted, but actively encouraged and
even helped by the Turkish authorities. Such a policy, in the
fifteenth and sixteenth centuries, demands an explanation.

In 1523, when the expulsion from Spain was still a matter

of living memory and concern, a Jewish historian named Eliyahu Capsali wrote a chronicle in which he said:

> Sultan Beyazid, the King of Turkey, heard of all the evil the King of Spain had done to the Jews, and that they were seeking a refuge, and he took pity on them. So he sent emissaries, and he made a proclamation throughout all his kingdom, and put it also in writing, that none of the governors of his cities was permitted to reject or expel the Jews, but that they must welcome them. And all of the people in all of the Kingdom welcomed the Jews, protecting them night and day. They were not abused, nor was any hurt done to them. Thousands and tens of thousands of those who had been expelled from Spain came to Turkey, and the land was filled with them. Then the Jewish communities of Turkey did countless great deeds of charity, and spent money like water to ranson captives.[13]

This last reference is to the numerous Jews who were captured by corsairs or simply seized and held for ransom by the captains or crews of the ships on which they had taken passage. Part of the text cited is a direct quotation from the Bible, from Ezra 1:1. The quotation refers to the Persian king Cyrus, in the Hebrew Bible the paradigm of the righteous and benevolent gentile ruler, who brings salvation and protection to the Jewish people.

Capsali was neither a Spanish Jew nor an Ottoman subject. Rather, he was a member of the old Jewish community of Crete, Greek by language and culture and a subject of the Venetian republic by political allegiance. He had no pretensions to historical scholarship; on the contrary, he seems in many ways to have been a rather simple and ignorant man. Historians, as we all know, commonly get their information from other historians. Capsali did not do that. As he explains in his book, he went around asking people for news of what had happened. "In our island," he says, "many merchants and many seamen come to our seaports from all over the world, and by talking to them it is possible to know what is going on."[14] This gives Capsali's narrative a special interest. Unlike most other historians of the time, what he provides is not a more or less official narrative of events, but a compendium of the gossip of Mediterranean seaports. He thus offers a reflection not necessarily of what actually happened, but of how the events of the time were perceived and felt by those who were most concerned. And that, surely, is at least equally important.

In particular, Capsali reflects the immense sense of tragedy, particularly among the Jews of Christendom, who felt the more threatened as the power of Christendom was visibly growing in the world. There was a need for relief, for a refuge, for somewhere to go, and that somewhere was found.

Capsali attributes the welcome given to the Jews to compassion; that is, the sultan took pity on them. His is obviously a rather mythic picture, and there is no evidence to show that Sultan Beyazid took a personal interest or was moved by such considerations. But as so often in such matters, a contemporary popular image—and on this, Capsali's perception is confirmed by other, more sophisticated Jewish writers—reflects a genuine historic truth.

Compassion is a noble sentiment and may suffice as the explanation of individual acts on particular occasions. It does not suffice as the explanation of a state policy pursued over a long period by successive generations of rulers and administrators. The historian will inevitably look for other explanations, if not in place of, at least in addition to, human compassion. In a search of human history, it would be difficult, if not impossible, to find major state policies determined by compassion, and such a determination does not seem likely in the circumstances of the present time. It is in the Turkish sources, particularly the documents, rather than in the Jewish sources that we may find an answer.

A few years ago, during a brief visit to Budapest, I found myself reading the tourist literature that the management had thoughtfully provided in my hotel room. Among other papers, there was a leaflet informing me that if I stepped out the front door of the hotel and walked a couple of hundred yards to the right, I would find the ruins of a synagogue and some Hebrew inscriptions. Acting on this instruction, I found some ruins, which might well have been a synagogue, and some unmistakable Hebrew inscriptions. The curious thing about these Hebrew inscriptions in the fortress of Buda was that the Hebrew script and style were those of the Sephardic Jews of Turkey, not of the Ashkenazic Jews of central Europe. What, one might ask, were Ottoman Sephardic Jews doing in Buda in the seventeenth century?

The history of central Europe in that period provides the outline, and documents in the Turkish archives provide the details, of an answer to this question. When the Turks conquered part of Hungary and set up an administration at the beginning of the sixteenth century, they brought their Jews with them and invited Hungarian Jews to go to Turkey. Then when they left in 1686, they took their Jews away with them. There are records of imperial orders to protect them, ensure their safe departure, and resettle them in suitable places in the Ottoman lands after the withdrawal from Hungary.

Documents from the Turkish archives, supplemented by evidence from other sources, indicate the reason as well as the circumstances. Let me illustrate with another personal experience. More than forty years ago, during my first incursion into the imperial Ottoman archives, I was going through the registers of imperial orders and came across a document dated 8 October 1576 and addressed to the governor of Safed in the Holy Land. The sultan had ordered "that a thousand Jews be registered from the town of Safed and its environs, and sent to the city of Famagusta in Cyprus." On receipt of this order, the governor was instructed "to register one thousand rich and prosperous Jews and send them, with their property and effects, and with their families, under an appropriate escort, to the said city."

Another order, dated 27 August 1577, is addressed to the qadis of Manṣūra and Quneiṭra, two places on the road from Safed to Damascus, and deals with the transfer of "500 Jewish families from among the rich and wealthy of the Jews of Safed" who are being transferred to Cyprus, in charge of an imperial pursuivant. Another order of the same date, to the governor general of Cyprus, informs him of these actions— "for the good of the said island"—and instructs him "to send each time a new galley from those at your disposal, bring them to the same island, and settle them in a suitable city." On this occasion, the Jews, who had no desire to leave the Holy Land for Cyprus, were able to persuade the Ottoman authorities to change their minds. An imperial order to the qadi of Safed, dated 23 May 1578, explains why. The annual poll tax paid by the Jews in Safed amounted to 1,500 gold pieces, and their extraordinary levies were paid in full to the imperial treasury. They paid 400 gold pieces every year as rent for a khan that was a pious foundation (*waqf*) for the benefit of the Dome of the Rock in Jerusalem and the Mosque of Hebron. The town's customs department received more than 10,000 gold pieces a year from the Jews. If they were transferred to Cyprus, the public revenue would lose all this money, and "the town of Safed will be on the verge of ruin. The Treasury of Damascus will suffer a great loss. . . . Their houses will also remain deserted; no buyer will be found [for them]. Their landed property will go for nothing. . . . Considerable loss and damage will result." The order to transfer the Jews to Cyprus was therefore countermanded: "They shall live at their present places and attend to their businesses. You shall not let [them] be troubled in the said matter."

An order of the following year to the governor of Cyprus indicated that he, for his part, still wanted his Jews. The governor had reported that he was holding a group of Jews who had stopped in Cyprus on their way from Salonika to Safed and requested permission to keep them and resettle them in Cyprus. Permission was granted.[15]

These orders were sent in the years following the Ottoman conquest of Cyprus from the Venetians in 1571 and are part of a pattern of such orders. The registers contain similar and contemporary orders from the sultan to various provincial governors in Anatolia, to send peasants and nomads to Cyprus to raise crops and tend herds. All this was part of a general Ottoman policy of transferring reliable elements—by invitation if possible, but by compulsion if not—into a newly conquered territory. The function of the Spanish Jews in Buda was the same as that of the Safed Jews in Cyprus and of others sent to newly conquered cities: to provide the Ottoman governor with an economically active and politically reliable element.

The Jews, unlike the Christians who formed the majority of the subject population, had no yearning for the lost glories of Byzantium, no desire whatever to be "liberated" from the Ottoman yoke. Many, perhaps most of them, were refugees from Christian countries. Indeed, of the many different ethnic and religious groups over whom the Ottoman sultans ruled, the Jews were the only volunteers.

The migration of European Jews to Turkey did not begin in 1492, nor did they come exclusively from Spain. Many, for good and obvious reasons, fled from the Italian lands under Spanish rule. They also came from other parts of Italy and even from central Europe. But the great majority of the communities listed in the Ottoman fiscal records of the sixteenth century bear the names of Spanish cities and provinces. The Spanish language that the Jews brought with them has survived to the present day and, by a strange paradox, was even adopted in many cities by the native Greek-speaking Byzantine Jews who had lived there before the Spanish exiles arrived.

If the pogroms in the Rhineland were a rehearsal for the Crusades, the expulsion of the Jews from Spain served a similar purpose for the expulsion of the Muslims.

During the centuries-long wars of the reconquest, when Spain was divided into Christian and Muslim principalities, the Christian princes were in general not less tolerant, and sometimes even more tolerant, than their Muslim colleagues and were willing to offer both Muslims and Jews living under their rule a protected but subordinate status similar to that accorded by Muslim rulers under the terms of the *dhimma,*[16] an arrangement by which they paid tribute or poll tax in return for the free exercise of their religion. Indeed, there were cases in medieval Spain of Jews fleeing from Muslim oppression to Christian tolerance. This was not, however, the usual pattern.

For the Jews, the reconquest initially meant no more than the exchange of one master for another. For the Muslims, the change was more difficult, since it was from domination to subordination. Great numbers were willing to accept this new status and to continue their lives under Christian rule. Those who accepted it were known in Spanish as *mudéjar,*[17] from the Arabic *mudajjan* or *ahl al-dajn*. The normal meaning of this word is "tame or domesticated animals," as opposed to "wild animals." Its use, by those Muslims who had voluntarily left their conquered homes or whose homes had not yet been conquered, to designate those who remained under Christian rule was an expression of contempt.

But more than contempt was involved. As the *reconquista* gathered force, Muslim jurists began to debate whether it was indeed permissible for Muslims to remain under non-Muslim rule or whether they should not instead follow the example of the Prophet in his migration from Mecca to Medina and leave for some other country, where a Muslim government ruled and the Holy Law of Islam prevailed. Among the doctors of the Mālikī school, followed by most of the Muslims of North Africa and Spain, the dominant opinion was that they must indeed leave in such a situation.[18]

The famous Ibn Rushd—qadi and imam of the Great Mosque of Cordova and grandfather of the still more famous philosopher of the same name, known in Europe as Averroës—ruled that the duty of *hijra,* or migration from infidel rule, is for all time and that a Muslim must not remain where he is subject to the jurisdiction of unbelievers, but must leave and go to a Muslim land. Like the companions of the Prophet, who were permitted to return to Mecca after it was conquered, so those who migrated in this way would be permitted to return to their homeland when it was restored to Islam.

A later Moroccan jurist, Ahmad al-Wansharīsī, went even further. Posing a purely hypothetical question—may they stay if the Christian conquerors are just and tolerant?—he answered no: Muslims still must leave, since only under Muslim rule can they live a true Islamic life, and under a tolerant infidel regime the danger of apostasy would be greater.[19]

In fact, Christian tolerance ended with the completion of the reconquest. The change was all the more dramatic in that the final conquest of Granada was achieved not by assault but by a negotiated capitulation, in which the ruler and nobles of Granada ensured their own safety and also secured from the victorious Catholic monarchs a whole series of undertakings, whose purpose was to guarantee to the Muslims possession of their homes and property, the free exercise of their religion, exemption from any tax other than those prescribed by Islamic law, and the right to be judged by their own judges in accordance with their own laws. Significantly, the capitulation also states:

> Those Moors, both great and small, men and women . . . who may wish to go and live in Barbary, or to such other places as they see fit, may sell their property, whether it be real estate or goods and chattels, in any manner and to whomsoever they like, nor would such property be confiscated, either from the sellers or the purchasers.

A number of other provisions guarantee the "right" of the "Moors" to depart with their families and "all their possessions of any kind whatsoever except firearms."[20]

The Muslims were not at this stage expelled, but the elaborate and favorable conditions laid down for their voluntary departure in the capitulation of Granada as well as in some earlier, similar documents reveal with some clarity the preference and ultimate intention of the conquerors. Almost immediately, the upper classes left in large numbers for North Africa.

The arrival in 1498 of the celebrated archbishop of Toledo, Francisco Jiménez de Cisneros, brought a change of policy, and Muslims found themselves under increasing pressure to convert or depart. By 14 December 1499, these pressures drove the remaining Muslims to open revolt, which was finally put down by armed force in 1501. After the suppression, increasing numbers of Muslims chose either conversion or emigration, and in February 1502, the remaining Muslim population of the kingdom of Castile, which by then included Granada, was offered the same choice that had been offered ten years earlier to the Jews: baptism, exile, or death.

In the years that followed, this choice was imposed on the Muslims in Navarre and Aragon and Valencia and other parts of Spain. Those who left were obliged to abandon most of their property; those who stayed were suspected, often with good reason, of practicing their old religion in secret and of complicity with the Muslim powers. They were known as *moriscos*,[21] the Muslim equivalent of the crypto-Jewish *marranos*. Both groups, being nominal Christians, were subject to the jurisdiction of the Inquisition and were the recipients of its special attention.

Throughout the sixteenth century, friction continued between the royal and ecclesiastical authorities, on the one hand, and the *moriscos,* on the other. The *moriscos* were always suspect and were often subject to discriminatory legislation and taxation, which in effect treated them as if they were still *mudéjares.* Although their numbers were diminished by more or less voluntary emigration, they remained an important element in the population, and their grievances occasionally burst into open rebellion. In 1608/1609, a meeting of Spanish bishops refused a request to condemn the *moriscos* collectively as apostates and instead urged a new effort to bring about their genuine conversion.

Nevertheless, in 1609, ostensibly because of a conspiracy with enemy foreign powers, a decision was taken to expel all of them. The expulsion was completed by 1614. The *morisco* population at the time of the expulsion has been estimated at some 320,000, about 3 percent of the total population. Most of them went to North Africa; some traveled, by sea and overland via Europe, to the Ottoman lands.

Even the expulsion of the Jews and Muslims and the subsequent repression and extrusion of the *marranos* and *moriscos* did not allay Spanish concerns about the perceived threat that they offered. An elaborate apparatus was constructed—legislative, investigative, executive—to unmask and forestall any Jewish or Muslim influence and to detect and excise any taint of Jewish or Moorish blood and thus protect the Spanish state, church, and society from contamination.[22]

These concerns persisted for centuries and were sometimes reinforced by new, external dangers. In 1704, the Rock of Gibraltar was captured by Great Britain, a country with a more relaxed attitude toward ethnic and religious differences.

The reported arrival in Gibraltar of a small number of "Jews and Moors" caused new anxiety in Spain. In Article X of the Treaty of Utrecht of 1713, the government of Spain ceded sovereignty over Gibraltar to Britain but, despite British reluctance, was able to insist on an important proviso:

> Her Britannic Majesty, at the request of the Catholic King, does consent and agree that no leave shall be given, under any pretence whatsoever, either to Jews or Moors, to reside, or have their dwellings, in the said town of Gibraltar.

After some initial attempts at compliance, this undertaking was, in the true sense, "more honour'd in the breach than in the observance." As early as 1721, it was breached by implication. In the Anglo-Moroccan treaty signed in that year, Article VII sets forth the rights of Englishmen to travel and trade in Morocco and adds:

> The like usage and treatment the Subjects of the Emperor are to receive in the King of Great Britain's Dominions.

A supplementary agreement, signed in 1729, is explicit:

> All Moors or Jews, subject to the Emperor of Morocco, shall be allowed a free traffic, to buy or sell for 30 days in the City of Gibraltar, or the Island of Minorca, but not to reside in either place, but to depart with their effects, without let or molestation, to any part of the said Emperor of Morocco's Dominions.

Even this limitation was soon abandoned, and during the eighteenth century, despite repeated protests by the Spanish embassy in London, Jews, mostly Moroccan, many of them of Spanish origin, came to form between a quarter and a third of the resident civil population of the Rock.[23]

Both the Muslim and the Jewish exiles continued for generations to mourn their lost homeland, which they remembered by its Arabic and Hebrew names as Al-Andalus and Sepharad. For centuries, Muslim families of Andalusian origin remained a recognizable social group in North Africa, with proud memories of their origins in Spain. Ultimately and inevitably, they were absorbed into the general Muslim population, but the sense of loss remained. The reports of the Moroccan ambassadors, probably the only Muslims apart from captives who visited Spain until relatively modern times, reflect this nostalgia for the lost glories of Andalus and follow almost every mention of a place name with the invocation "May God speedily return it to Islam."

The Jewish exiles also in the course of time merged with their co-religionists in the countries in which they had found refuge, but only after tensions and squabbles with those whom in North Africa they called *berberiscos* and, in the east, *griegos*. In North Africa, the Spanish Jews were in time absorbed by the native North Africans. In Turkey and the Balkans, it was the Spanish Jews, more numerous and better educated, who imposed their language and culture on the native Greek-speaking Jews remaining from Byzantine times.[24]

The two greatest centers of Ottoman Sephardic Jewry were Istanbul and Salonika, in both of which large numbers of Jews, native and foreign, were settled by the Ottoman authorities as an act of imperial policy. After the conquest of Constantinople in 1453, the sultan set about the task of rebuilding and repopulating his new imperial capital. Jews

played an important role in his plans. As part of the recon-
struction, many Jews were compulsorily transferred from
various provinces of the Ottoman realm to the new capital.
And because there were not enough of these Jews, refugees
from outside were particularly welcome.

Even more remarkable was the case of Salonika (in Greek,
Thessaloniki). The early Turkish registers after the capture of
the city in 1430 show a mainly Greek population, with a
small but increasing number of Turks and a small and dimin-
ishing number of Catholics. As late as 1478, a Turkish register of households in the city shows no Jews. An undated
register of the time of Süleyman the Magnificent, about half a
century later, lists no fewer than twenty Jewish communities.
By the early seventeenth century, the number of communities
had risen to twenty-five. By this time, Jews were a clear majority of the total population of Salonika, which remained a
predominantly Jewish city until the early twentieth century.

For more than four centuries, Salonika was an economic
center in Ottoman Europe and a cultural center of Sephardic
Judaism. This center was created by those who, confronted
with the choice between conversion and exile, chose exile.
Their remote descendants in the twentieth century, when
Greece was conquered and occupied by the armies of the
Third Reich, were offered no such choice, and all but a few of
them perished in the death camps of Nazi Germany. They
would surely have been glad to be offered the choices of
fifteenth-century Catholic Spain.

After the destruction of Spanish Judaism and Islam, there
were some in both communities who tried to preserve the
memories of past achievements and sufferings. Jewish historians in France, Italy, and Turkey wrote martyrologies, in

which Spain and Portugal made a predominant, though not an exclusive, appearance. Other Jewish scholars preserved and in due course printed the writings of the many great Jewish poets and philosophers who had flourished in medieval Spain. In the seventeenth century, a North African Muslim scholar and man of letters, Aḥmad ibn Muḥammad al-Maqqarī, devoted himself to the study of Muslim Spain and produced a monumental work on its history and literature.[25]

But gradually even these memories faded and were lost, until the early nineteenth century, when the Romantic revival in western Europe evoked a new interest in medieval Spain, which in time included both its Jews and its Muslims. This was the time when Victor Hugo's play *Hernani,* performed in 1830, and Theophile Gautier's *Voyage en Espagne,* published in 1845, set the fashion in Paris, and Washington Irving's two books, *The Conquest of Granada* and *Tales of the Alhambra,* won immense popularity in the English-speaking world. In 1833, a Frenchman named Louis Viardot published his *Essai sur l'histoire des Arabes et des Mores d'Espagne*. A Turkish translation by one of the leading Turkish intellectuals of the time followed some years later, and a new cult of Spanish Islam was launched in the Muslim East.[26] In London, a Sephardic Jew named Elias Haim Lindo published his *History of the Jews in Spain and Portugal* in 1846. Lindo was a cousin of Benjamin Disraeli and clearly influenced the Sephardic pride and the mythic view of the role and place of Jews in Muslim Spain reflected in his more famous cousin's novels.

Jews and Muslims left Spain as exiles, and most of them

went south and east seeking refuge. But at the same time, great numbers of Christians, new as well as old, were leaving Spain and traveling not east but west, not as refugees but as conquerors, colonists, and, almost incidentally, discoverers of a new world.

CHAPTER 3
Discovery

*I*N THE LATE THIRTEENTH CENTURY, when the reconquest was well under way but far from completed, Ramon Llull, one of the principal theoreticians of the Christian war against Islam, remarked that when the recovery of Spain was completed, it would be necessary to carry the war beyond the Strait of Gibraltar to the other side.[1] He was not alone in this opinion. Similar perceptions and ambitions obviously inspired many of the Spanish and Portuguese rulers and commanders, not only during but also after the liberation of their national soil. Similar thoughts obviously occurred to the Muscovites, who, after freeing Moscow from Tatar rule, pursued their fallen masters into Tartary.

The same underlying assumption inspired all these policies: that by driving the Muslims out of Iberia and Russia, they had won a great battle, but they had not won the war, and that the long, drawn-out struggle between Christianity and Islam continued, on a vaster scale. Both Spaniards and Portuguese saw themselves as continuing the same struggle against the same enemy, and it was natural that when they encountered Muslims as far away as Ceylon and the Philippines, they called them Moors.

There was, of course, another, more practical, aspect to this new phase of what Gibbon called "the great debate." When the Portuguese explorer Vasco da Gama, arriving in India a few years after Columbus had sailed, remarked that he had come "in search of Christians and spices," he was expressing perfectly this double aspect of the voyages of discovery. On the one hand, it was a strategic move in a religious war of global dimensions; on the other, it was a commercial ploy designed to cut out the middleman and go straight to the producer.

This was in a sense a conflict that antedated both Islam and Christendom. In the days when the rival empires of East and West were Rome and Persia and when Persia lay astride Rome's routes to the East, the Romans tried to establish direct commercial links with China, the source of silk; with India, of hardwoods; with Southeast Asia, of spices. They did this by exploring the two outer routes that lay beyond both sets of imperial frontiers, venturing northward into the Eurasian steppes and southward into the Arabian Peninsula. The Christianization of Rome's successors and the Islamization of the Eastern empire added a religious, and therefore also a new military and strategic, dimension to this ancient rivalry.

There were good, practical reasons for the victors in the reconquests, at both ends of Europe, to pursue their retreat-

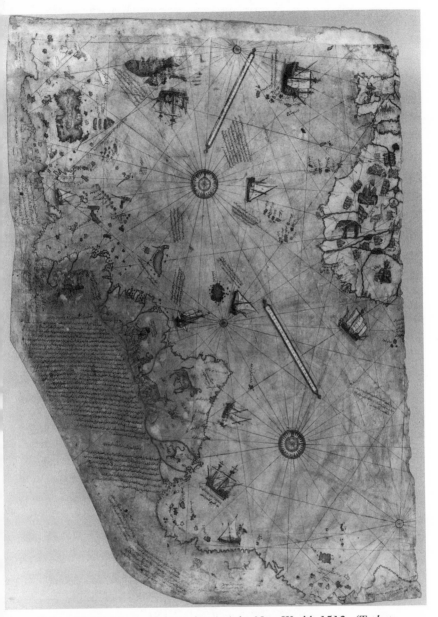

Part of Piri Reis's world map, showing the New World, 1513. *(Topkapı Palace Museum Library, Istanbul. Courtesy of the Ministry of Culture of the Turkish Republic)*

Christopher Columbus at the court of King Ferdinand. Miniature from a Turkish manuscript of the *Tarih-i Hind-i Garbi* (History of the West Indies), 1583–1584. *(Beyazid Library, Istanbul. Courtesy of the Ministry of Culture of the Turkish Republic)*

Portuguese Jew, wearing a prayer shawl and holding a Torah scroll,
Holland, seventeenth century. *(Musée National du Moyen Âge, Paris.
Courtesy of the Réunion des Musées Nationaux de France)*

Costume of the Barbary Jew Hawker in Gibraltar

North African Jewish peddler in Gibraltar, eighteenth century. *(Courtesy of the Alfred Rubens Collection, London)*

European, probably Spanish or Portuguese, visitor to Iran. Wall painting in the Chihil Sutun pavilion in Isfahan, late sixteenth century, rebuilt 1706. *(Photograph by the author)*

یحیی ذیمان رافرمان داد آنروز دیگر یک بریک دلیل حود درشیشه کنند و پیاورند و سکره بود در میان نیان

Jesuit priest visiting an elderly Indian ruler, late sixteenth century. (*Courtesy of the L. A. Mayer Memorial Institute for Islamic Art, Jerusalem*)

Young Portuguese woman, India, seventeenth century. *(Courtesy of the L. A. Mayer Memorial Institute for Islamic Art, Jerusalem)*

Woman of the New World. Miniature from a Turkish manuscript of the *Zenan-name* (Book of Women) of Fazil Bey, late eighteenth century. *(British Library, London. By permission of The British Library)*

ing enemies. There was the obvious tactical need to prevent them from regrouping and to forestall a counterattack. There was the growing desire to break free from the Muslim pincers that had gripped Europe from east and south since the high Middle Ages. Beyond these there was the more ambitious aim, by no means new, to outflank the power of Islam by finding co-religionists, business partners, and perhaps even allies in the remote lands beyond the eastern and southern limits of Muslim power.

The recurring dream of an anti-Muslim second front haunted both statesmen and writers during the European Middle Ages. It found expression, on the one hand, in attempts to cultivate relations with the Mongols, the Ethiopians, and even the Shi'ite Persians and, on the other, in the protean myth of Prester John,[2] the variously located great Christian king of the East who one day would join forces with Europe to bring about the final defeat of the Islamic enemy. The first mention of Prester John in a Western source appears to be that of the German ecclesiastical historian Otto, bishop of Freising. In 1145, Otto tells us, he had a meeting in Viterbo in Italy with the bishop of Byblos, or Jbeil, in Lebanon. Among other things, the bishop told him:

> Not many years ago, a certain Johannes, a king and a priest, living in the Far East [*in extremo Oriente*] beyond Persia and Armenia, who like all his people was a Christian though a Nestorian, made war on . . . the kings of the Persians and Medes . . . and stormed the capital of their kingdom. . . . At last, presbyter Johannes—for so they are in the habit of calling him—was victorious, putting the Persians to flight with most bloodthirsty slaughter.[3]

After this victory, Bishop Otto relates, on the authority of his colleague from Lebanon, Johannes "had advanced to the help of Jerusalem" but had been unable to take his army across the river Tigris and had therefore turned north, where he and his army encountered extremely cold weather and were obliged to return home.

Modern historians have agreed that this curious story reflects a real event—the Battle of the Katvan Steppe, east of Samarkand, where in September 1141 the Seljuq Great Sultan Sanjar was crushingly defeated by the Karakhitay, an East Asian people related to the Mongols. Like most of the steppe peoples, the Karakhitay were tolerant and even impartial in religious matters, and under their aegis, the Nestorian Christian church of Inner Asia enjoyed a new revival and a great expansion of missionary activity.

Despite Otto's caution—he holds out no hope of salvation from the East—news of a major Muslim defeat at the hands of a new power, which, if not Christian, gave shelter and encouragement to Christian churches, did not fail to arouse excitement and expectations in Europe. This excitement and these expectations were heightened in the following century, when news was received of the great conquests by the Mongol hordes and the destruction by them of the Baghdad caliphate. Not only were there Nestorians—schismatics but still Christians—among the Mongols, but some of them had actually risen to positions of command in the Mongol state. These hopes seemed to be fulfilled in a dramatic moment in March 1260, when, after the Mongol capture of Damascus, three Christian princes rode abreast through the streets of the conquered Muslim city: the Mongol commander Kitbuga Noyon, a Turk by origin and a Nestorian

Christian by religion; the Frankish crusader Bohemond VI of Antioch; and his father-in-law, Hethoum, of the renascent Franco-Armenian kingdom in Cilicia.

But these hopes were soon dashed. Damascus was recaptured for Islam: Bohemond was excommunicated by the pope; and the local Christian population of Damascus paid a price for this brief extravaganza. As for the Mongols themselves, despite all the efforts of Christian preachers to convert them and of Christian diplomats to recruit them, they opted for Islam. Their conversion to the dominant faith of the region they had conquered dealt a heavy blow to the hopes of Christians and gave a new strength and vigor to their Muslim opponents.

This did not, however, end the hope or the quest. As late as the fifteenth century, attempts were still being made, this time in Africa, not Asia, to find a Christian ally that would open a second front. In both Arab and European sources, there are occasional references, tantalizingly brief, to exchanges between Europe and Ethiopia with a view to common action. These exchanges were, of necessity, secret, since their only lines of communication ran through Muslim-held territory. The Egyptian historians tell a curious story of the arrest, trial, and execution in 1428 of a Persian merchant, accused of having acted as intermediary between Ethiopia and various "Frankish" rulers and of having tried to organize a combined assault from both sides, "to overwhelm the Muslims and put an end to Islam."[4] Not surprisingly, these efforts came to nothing, but the grand design of outflanking Islam and opening a second front continued to inspire would-be crusaders, explorers, and those who tried to combine both roles.

Vasco da Gama, not surprisingly, put spices second to Christians in his statement of objectives. Others might reverse the order and take spices as the symbol of a larger category. There were, of course, substantial economic motives. Europe was beginning to run out of precious metals, and there was great interest in the famed gold of Africa, in the gold caravans that traveled across the Sahara. There was a growing appetite for sugar, another import from the Islamic Middle East. The transplantation and cultivation of sugar in the Atlantic islands, the Canaries and the Azores, were an important stage in the Atlantic discoveries. And there was, of course, the mounting impatience of European traders and consumers with the exorbitant demands of Middle Eastern middlemen and tax collectors, and the desire to go directly to the source of supply of the much-coveted spices of South and Southeast Asia.

In this larger historical context, it is easier to understand the beginnings of European expansion in the late fifteenth century. It was natural that the Spaniards and the Portuguese, having liberated their homelands from Muslim rule, should pursue their former rulers across the Strait of Gibraltar, if only to make sure that they would not return. In the same way, the victorious Russians, having shaken off the Tatar yoke in Muscovy, naturally and inevitably pursued the Tatars to their own home bases, acquired in earlier conquests. In the moment of victory after a long and bitter holy war, one does not arbitrarily halt the advance of one's victorious armies pursuing the defeated foe, because they have reached some theoretical line between imaginary entities. For Russians and Iberians alike, Europe was Christendom, and its frontier, in the east as in the south, was Islam. And naturally, the frontier between two rival, universal religions and civilizations had no geographical fixity, but would move with the tide of battle until—in the common conviction of both—it ended with a final victory of the true faith. This was also the view that the Muslims held and had conveyed to Europe—which indeed they had held since the seventh century, when they had inaugurated the long sequence of attack and counterattack, *jihād* and crusade, conquest and reconquest, and which, even in the Age of Discovery, was still represented by the Turkish armies on the Danube.

The imaginary entities to which I refer are, of course, Europe, Asia, and Africa, the three continents into which, by ancient tradition, the Old World was divided. A clear and simple description is provided in the *Natural History* of Pliny:

> The whole circuit of the earth is divided into three parts, Europe, Asia and Africa. The starting point is in the west, at the Straits of Gibraltar, where the Atlantic Ocean bursts in and spreads out into the inland sea. On the right as you enter from the ocean is Africa, and on the left Europe, with Asia between them; the boundaries are the river Don and the river Nile.[5]

This description clearly comes from a society whose world is defined by the Mediterranean, its shores, and its hinterland. It is derived, like most Roman science and philosophy, from Greek writings, many of which are still extant. In these, the southern continent is not Africa but Libya; the other two names are the same. The name Libya, in various forms, occurs in both ancient Egyptian and biblical Hebrew texts and appears to denote a people and region west of Egypt. Its use as the name of the southern continent is known only in Greek.

All the ancient texts in which the three names—Europe, Asia, and Africa—occur are, without exception, Latin or Greek, that is, European. In the by now considerable body of writings that have come down to us from the ancient civilizations of Asia and Africa, there is not so far a single occurrence of these names. This would appear to confirm that even though the information on which it was based may have derived at least in part from Phoenician explorations, the geography of a tricontinental world was a Greek concept. Europe was what the Greeks—and later the Romans and others—called their own homeland. Asia and Africa were what they called the lands of their neighbors, but that was not what these neighbors called themselves. Europe was itself, self-defined in the successive phases of Hellenismos, Romanitas, and Christianity. Asia and Africa were designations of the Other—geographically expressed equivalents of such ethnic, cultural, and religious designations as barbarian and gentile. The barbarians did not, of course, call themselves barbarians, nor did the gentiles call themselves gentiles until they were converted to Christianity and taught to see themselves as such. In this perspective, Asia simply meant not-Europe East, and Africa or Libya meant not-Europe South. Some ancient philosophers, among them Aristotle, went a step further and equated Europe with freedom, independence, and the rule of law, and its neighbors with arbitrary tyranny and slavish submission. This view is still sometimes expressed in the councils of the European Union.

Needless to say, the inhabitants of Asia and Africa did not share this perception, and as far as the evidence goes, they were as unaware of being Asians and Africans as the inhabitants of pre-Columbian America were unaware of being

Americans. They first became aware of this classification when it was brought to them—and at some times and in some places imposed on them—by Europeans.

It is therefore not surprising that while Europe, despite its many nations and languages, is a single coherent entity—a genuine community of culture, values, religion, institutions, science, arts, and even music—and is, or was until recently, inhabited by people of a single race, Africa and still more Asia show an immense diversity. The differences between Sicilians and Swedes, between Poles and Portuguese may loom large in Europe, but they are insignificant compared with the differences between India and Japan, between Iran and China, between Egypt and Zimbabwe. Europe became Europe. Europe discovered and in a sense created America, since every polity in the Western Hemisphere, from the Arctic to the Antarctic, was founded on a European model and expresses itself in a European language. Europe neither created nor discovered Asia and Africa. It invented them, and it is a supreme irony of our time that in a wave of revolt against Eurocentrism, so many non-Europeans have adopted this ultimately Eurocentric view of the world and defined themselves by the identity that Europeans imposed on them.

Historically, culturally, and even racially, there was no real barrier or dividing line between Europe and its eastern and southern neighbors; indeed, those parts of Asia and Africa that the Greeks and Romans knew and named had far more in common with Europe than with the more distant regions of the two continents. The Mediterranean was always far less of a barrier than the Sahara or than the mountains and steppes that mark the northern and eastern limits of Iran. Compared with the great religions of India and China, Islam and Christianity are twin offspring of the same progenitors. Compared with the races of southern and eastern Asia and of central and southern Africa, the peoples of all the shores of the Mediterranean clearly share a heritage. There is, in fact, no clear racial dividing line between white and nonwhite at any of the extremities of Europe—rather, a series of gradations, for example, from north to south, from blond to brunet to dark to swarthy to brown to black.

For the Greeks, Asia originally meant the eastern shore of the Aegean Sea, of which the western shore was Europe. Then, as a vaster, more remote Asia loomed on their horizon, it was renamed Microasia, Asia Minor, to distinguish it from the more distant realms of the Persians, the Indians, and the Chinese. In much the same way, the familiar East, which gave us the Eastern Question, became the Near and then the Middle East when the problems of a Far East attracted Western attention.

Under Roman and then Byzantine rule, Asia was the name of a province. It remained in use until the Islamic conquest, after which it disappeared from local usage. Neither the Arabs who conquered Syria nor the Turks who conquered Asia Minor retained the use of this name.

For the Romans, Africa was the name of the Mediterranean shore that faced them in the south: home of Carthage, the one enemy power that seriously threatened the might and for a while even the survival of Rome. Scipio, the Roman who was most acutely and vocally aware of this danger, was known as Africanus, the African, not because he came from Africa, but because he conquered it. In the Roman Empire, Africa became an imperial province corresponding roughly to present-day Tunisia plus part of eastern Algeria and western

Libya. Unlike Asia, the name Africa survived the Arab conquest of the seventh and eighth centuries and was retained by the new masters as a geographical and, for a while, administrative expression to designate the same area. It did not include Egypt or Morocco, still less the countries to the south, which were known in Arabic as Bilād al-Sūdān, the lands of the blacks. Originally applied to the whole sub-Saharan region from the Atlantic to the Red Sea and Indian Ocean, this name has in modern times been restricted to two countries in Northeast Africa. In local usage in the Nilotic Sudan, the brown northerners are called red, while the black southerners are called blue.

How, then, did the Muslims divide the world? They did so in two ways: one geographical, the other at once religious and political. In the geographical literature in Arabic, Persian, Turkish, and other Muslim languages, the world is divided into "climates," in Arabic *iqlīm*.[6] The classical Muslim authors did not, however, commit our modern error of ascribing racial or cultural or even historical attributes to these divisions. Their "climates" are purely geographical and have no religious, cultural, ethnic, or political connotation whatsoever. For Muslims, the important and operative division of the world was between the lands where Islam prevailed and those that had not yet been brought into the fold. In their advance in the Asian and African continents, the Muslims encountered no major military power and no serious religious rival. In Europe, they met both, and it is therefore not surprising that the theory as well as the practice of the *jihād,* the holy war for Islam, should have been mainly concerned with the struggle against Christendom in its various European theaters.

There is a curious parallelism in the mutual perceptions of the two. Each saw the other as a major rival, yet was unwilling to admit this. Rather than speak of their enemies as Muslims, Christians preferred to use ethnic terms and referred to them in different parts of Europe as Moors or Saracens, Turks or Tatars. In the same way, when Muslims spoke of the inhabitants of Europe or even of the Crusaders who had come from Europe, they named them as Greeks or Romans, Franks or Slavs. When some religious designation was required, both used the same term, *infidel*. By this, both Christians and Muslims meant the same thing, differing only in its application.

Europe and its sisters Asia and Africa reappear in European writings at the time of the Renaissance and the consequent revival of Greek learning. Rediscovered ancient classical texts gave Europeans a new way of looking not only at the past, but also at the major events of their own time. Once again, as in antiquity, Greece was being invaded by the vast armies of a great king in the east. In the fresh perspective of the new learning, this was no longer seen primarily as a continuation of the old struggle between Islam and Christendom, but as a renewal of a more ancient struggle—the defense of free European Hellas against an Asiatic tyrant and his barbaric hordes. This equation, however inappropriate, continued to affect European perceptions of events in the eastern Mediterranean for a long time to come. There was no comparable threat from Africa, but here, too, the old word was given new meanings derived from ancient texts and recent explorations.

The parallel discoveries—of antiquity by scholars, of the world by travelers, of their own language by humanists—and the weakening grip of religious authority on the mind of Europe all combined to encourage a new perspective in which religion was no longer the primary, still less the sole, definition of identity, otherness, and conflict. The growth of knowledge, the establishment of contact with more remote societies and cultures on the far side of the Islamic world, and, perhaps most important of all, the circumnavigation of the globe and the discovery of a new world with ancient civilizations equally unknown to the Greeks, the Bible, and the Qur'ān—all helped in time to encourage a more objectively geographical classification of the world. The discovery of a fourth continent lent new meaning and validity to the old established three.

In a sense, the appearance and acceptance of America ensured and confirmed the identity of Asia and Africa, since these too were European notions. It was above all the discovery of America, as part of the European discovery of the world, that ensured the triumph of Europe over its rivals, especially Islam, and the consequent universal acceptance of European notions and categories. After some initial curiosity, Muslim interest in what they called "the New World"—the name America rarely occurs—was minimal. The most remarkable document was a map prepared in 1513 by the great Turkish navigator and cartographer Piri Reis that includes information drawn from Columbus's own lost map and apparently supplied by a Spanish slave in the Turkish service. In 1517, Piri Reis presented the map in Cairo to Sultan Selim I, the conqueror of Egypt. It was deposited in the Topkapı Palace and forgotten until it was noticed by a German scholar in 1929.[7] The earliest, and for a long time the only, Muslim work on America was the *Tarih-i Hind-i Garbi,* a Turkish account of the West Indies, written in the mid-sixteenth century and printed in 1729 at the first Ottoman Turkish printing press. The unnamed author speaks in some detail of the discovery and conquest of the New World; describes its flora, fauna, and inhabitants; and, naturally, expresses the hope that this well-favored land would in due course be illuminated by the rites of Islam and added to the Ottoman realms.[8] Arabic and Turkish writings in the seventeenth and eighteenth centuries show some slight awareness of the American colonies of their European neighbors, and a late-eighteenth-century Turkish book on the women of the world even includes a fanciful description in verse of the (native) American woman, with a picture. A Chaldean (Uniate) Christian priest from Mosul traveled in Spanish America between 1675 and 1683 and wrote an account, in Arabic, of his travels.[9]

I do not know at what stage or by what processes the descendants of the Aztecs and the Incas came to understand and accept their identity as Americans. In Asia and Africa, the parallel processes can be dated with fair precision. They began with the advent of European domination or at least influence, when Asians and Africans studied the hitherto despised languages of the Western infidels and barbarians and began to learn geography from European textbooks. At the beginning of the nineteenth century, the Turks—for hundreds of years the front line of Islam against Christendom—still spoke of the countries beyond their western frontier as the lands of the Franks, the Christians, or the infidels, but not as Europe. During the first half of the nineteenth century, the continental classification and the nomenclature that went with it gradually entered local usage. The name Europe, in the form Arūfa, had made a fleeting appearance in early Arabic geographical works based on Greek originals and had quickly disappeared and been forgotten. In the nineteenth century, it reappeared, and the Arabic Urubba, Turkish Avrupa, and Persian Urupa passed into common use. In countries under direct European rule, like Russian Central Asia, French North Africa, British India, the Dutch East Indies, and most of Africa, the impact of European geographical knowledge and instruction was more direct and more immediate.

The adoption of this European scheme of things naturally entailed the recognition of Asia and Africa as continents, though it did not, for some time at least, involve their acceptance as identities comparable with that of Europe. The notion of Asianism did not emerge until late in the nineteenth century and developed in the twentieth, principally as an anti-Western movement, sometimes to promote anti-imperialism, sometimes a rival Asian imperialism. Apparently by mutual consent, the term Asian in this political context did not include the Muslim peoples of the southwest. It still does not include them in current American usage. The notion of a common African identity embracing the different religions, languages, cultures, and races of the continent from the Mediterranean to the Cape of Good Hope was a much later development and did not really become significant until after the ending of European rule. If Africanism survives, this form of revolt against a Eurocentric universe will mark a final triumph of Eurocentrism in Africa.

The oceanic voyages of the European explorers around Africa to Asia, across the Atlantic to the Americas, created, for the first time in history, a new unity of all the continents, bringing all of them into contact with one another and preparing the way for a global interchange of foodstuffs and commodities, plants and domestic animals, knowledge and ideas.

There was also a negative side. These new intercontinental lines of communication also made possible an interchange of diseases between the Eastern and Western Hemispheres, sometimes leading to the emergence of virulent new strains calling for new diagnoses and remedies. These could be social as well as medical, such as the new strain of slavery added to the numerous and widespread slave institutions of both the Old and New Worlds.

Although it was known in medieval Europe, slavery was of minor importance there, far less significant in the social and economic life of Europe than in pre-Columbian America or in Muslim and non-Muslim Africa. The meeting of all these different cultures gave rise to a new variant, known as colonial slavery. The fertile lands of America offered both opportunity and challenge. The inventiveness and cupidity of Europe, learning from and drawing on the plantation systems and the slave trade of Africa and the Islamic world, found this answer. Colonial slavery and the seaborne slave trade became a major factor in the crisscrossing interchanges between the four shores of the Atlantic—western Europe, western Africa, North America, and South America.

But it was Europe, too, that first decided to set the slaves free: at home, then in the colonies, and finally in all the world. Western technology made slavery unnecessary; Western ideas made it intolerable. There have been many slaveries, but there has been only one abolition, which eventually shattered even the rooted and ramified slave systems of the Old World.

In all this, as in much else, the discovery of America, for better or for worse, was a turning point in human history and an essential part of the transition to a modernity that began in Europe and was carried all over the world by European discoverers, conquerors, missionaries, colonists, and, let us add, refugees. Far more than the contemporary conquest of Granada, it ensured, in the long run, the triumph of Europe over its enemies. The mines of the New World gave European Christendom gold and silver to finance its trade, its wars, and its inventions. The fields and plantations of the Americas gave it new resources and commodities and enabled Europeans, for the first time, to trade with the Muslims and others as equals and, ultimately, as superiors. And the very encounter with strange lands and peoples, unknown to history and scripture alike, contributed mightily to the breaking of intellectual molds and the freeing of the human mind and spirit. One might say that, figuratively, Columbus, too, was looking for Christians and spices and also for Prester John. One might add that, symbolically, he found them all and much more.

Why, then, did the peoples of Europe embark on this vast expansion and, by means of conquest, conversion, and colonization, attempt to create a Eurocentric world? Was it, as some believe, because of some deep-seated, perhaps hereditary vice—some profound moral flaw? The question is unanswerable because it is wrongly posed. In setting out to conquer, subjugate, and despoil other peoples, the Europeans were merely following the example set them by their neighbors and predecessors and, indeed, conforming to the common practice of mankind. In particular, their attack on the neighboring lands of Islam in Africa and Asia was a clear case of be-done-by-as-you-did. The interesting questions are not why they tried, but why they succeeded and why, having succeeded, they repented of their success as of a sin. The success was unique in modern times; the repentance, in all of recorded history.

The attempt was due to their common humanity; the success, to some special qualities inherent in the civilization of Europe and its daughters and deficient or lacking in others. No doubt the Europeans had the mixture of appetite, ferocity, smugness, and sense of mission that are essential to the imperial mood and that they shared with their various imperial predecessors. But they also had something else, which both the former conquerors and those whom they conquered and then relinquished might find it useful to examine. Among both those who acclaim and those who denounce Columbus's voyage of discovery, there are some, it would seem, who base their judgment on the history not of America but of the United States of America, for which Columbus is seen as ultimately responsible. In our own time, this conceit has acquired a kind of absurd plausibility.

In this century, important changes have taken place in the configuration of European civilization. Under leaders profoundly hostile to European values, Russia withdrew from Europe and after the Second World War forced much of eastern Europe to do the same. Now, painfully and with difficulty, they are trying to return. Western Europe, weakened and exhausted by two devastating world wars and the loss of empire, yielded to America, where a new civilization had arisen, European in its origins and many of its values, yet profoundly different in others, giving new leadership to humankind in almost every significant field of human endeavor. It has now become customary to designate this larger civilization of which Europe is the source and America the leader as "the West." In addition to its obvious geographical denotation, this term has had two overlapping but somewhat different meanings. The one was a military alliance against the Soviet Union, including some countries that shared few or none of the basic values of the West and were linked to it only by strategic necessity. The other was a comity of like-minded nations that shared certain basic values concerning freedom and decency and human rights and included neutrals that wished no part of the anti-Soviet alliance. Today, with the collapse of the Soviet Union and the end of the Cold War, a new, larger, and perhaps even greater Europe may emerge. The West, no longer hemmed in by military needs and constraints, may also aspire to yet greater achievements. But for the moment, among the Europeans and among their children and disciples, especially in North America, the mood of greed and self-confidence has given way to one of satiety, guilt, and doubt.

Doubt is good and, indeed, is one of the mainsprings of Western civilization. It undermines the certitudes that in other civilizations and in earlier stages of our own have fettered thought, weakened or ended tolerance, and prevented the emergence of that cooperation of opponents that we call democracy. It leads to questioning and thus to discoveries, new achievements, and new knowledge, including the knowledge of other civilizations. Guilt in the modern sense—not a legal decision, but a mental condition—is corrosive and destructive and is an extreme form of the arrogant self-indulgence that is the deepest and most characteristic flaw of our Western civilization. To claim responsibility for all the ills of the world is a new version of the "white man's burden," no less flattering to ourselves, no less condescending to others, than that of our imperial predecessors, who with equal vanity and absurdity claimed to be the source of all that is good.

A word that is much heard nowadays is *multiculturalism,* expressing the idea that what is described as the exclusively Eurocentric character of our education and culture should be ended and replaced by one based on many cultures. The idea of multiculturalism is in itself excellent and very much in the great tradition of Europe and the West. It was, after all, the West that first, and for long alone, undertook the scholarly study of alien cultures with curiosity, interest, and respect and even—by exhuming buried civilizations and deciphering forgotten scripts—made a substantial contribution to their recognition and understanding of themselves.

In any case, European civilization is by no means exclusively European. Like every other culture known to history, it was enriched by contributions and influences from its predecessors, notably in Egypt and the lands of the Fertile Crescent. The study of these other cultures at the present day is valuable and important first for what they are in themselves, and second because only by studying others can we achieve a more profound and more realistic understanding of ourselves. It may seem quixotic to plead for the study of exotic and difficult languages at a time when our students seem unwilling to learn even the languages of our immediate neighbors and are having increasing difficulties with their own. But I must, as an Arabist, put in a word for Arabic, one of the great classical languages of human civilization and the vehicle of an immensely rich literary, religious, philosophical, and scientific culture.

Multiculturalism becomes dangerous and demeaning to all cultures when it presents an idealized and sometimes invented version of other cultures and contrasts them with a demonized parody of the West. Certainly, our civilization has many profound flaws, some of them part of our common humanity, others distinctively our own. To many of the charges leveled against us, we have no choice but to plead guilty—as human beings. We of the West have indeed been guilty of arrogance and domination, of aggression and spoliation, of subjugation and enslavement, of murder and rapine, although we might argue in mitigation that it has been a long time since we accepted these things as permissible and even longer since we regarded them as pleasing to God and as ordained by divine law.

By the mid-twentieth century, the New World, whose new history began in 1492, had become the unquestioned leader of Western civilization and its principal defender against those enemies, both internal and external, that had sought to destroy it. Even today, though that leadership is criticized by some, challenged by others, denounced and rejected by others again, it still faces no serious competitor, no viable alternative. Even those who daily cry "Death to America" may perhaps have the will, but surely lack the strength, to achieve that purpose. They cannot kill, but they might lend a hand in an assisted suicide if certain trends discernible in American societies reach that point. And if that happens, then in all probability not they, but others, would claim the inheritance.

But what is this inheritance, and is it worth preserving or claiming?

All cultures have their achievements—their art and music, philosophy and science, literature and lifestyles, and other contributions to the advancement of humankind—and there can be no doubt that knowledge of these would benefit us and enrich our lives. The recognition of this infinite human variety and the need to study and learn from it is perhaps one of the West's most creative innovations. For it is only the West that has developed this curiosity about other cultures, this willingness to learn their languages and study their ways, to appreciate and to respect their achievements. The other great civilizations known to history have all, without exception, seen themselves as self-sufficient and regarded the outsider, or even the subculture or low-status insider, with contempt, as barbarians, gentiles, untouchables, unbelievers, foreign devils, and other more intimate, less formal terms of opprobrium. Only under the pressure of conquest and domination did they make an effort to learn the languages of other civilizations and, in self-defense, try to understand the ideas and the ways of the current rulers of their world. They would learn, in other words, from those whom they were constrained to recognize as their masters in either sense or both, as rulers or as teachers. The special combination of unconstrained curiosity concerning the Other and unforced respect for his otherness remains a distinctive feature of Western and Westernized cultures and is still regarded with bafflement and anger by those who neither share nor understand it.

We of the West have often failed catastrophically in respect for those who differ from us, as our dismal record of wars and persecutions may attest. But it is something for which we have striven as an ideal and in which we have achieved some success, both in practicing it ourselves and in imparting it to others. It is surely significant that so many refugees fleeing from Vietnam made for the crowded island of Hong Kong—the one spot in all East Asia where a Western government still ruled and where they could therefore count on the certainty of public scrutiny and concern and the consequent hope, however slight, for help.

Imperialism, sexism, and *racism* are words of Western coinage, not because the West invented these evils, which are alas universal, but because the West recognized and named and condemned them as evils and struggled mightily—and not entirely in vain—to weaken their hold and to help their victims. If, to borrow a phrase, Western culture does indeed "go," imperialism, sexism, and racism will not go with it. More likely casualties will be the freedom to denounce them and the effort to end them.

It may be that Western culture will indeed go: The lack of conviction of many of those who should be its defenders and the passionate intensity of its accusers may well join to complete its destruction. But if it does go, the men and women of all the continents will thereby be impoverished and endangered.

Notes

I

1. There is a considerable literature on medieval European perceptions of Islam and the Muslims, as expressed in literature and scholarship. For some examples, see R. W. Southern, *Western Views of Islam in the Middle Ages* (Cambridge, Mass., 1962); Aldobrandino Malvezzi, *L'Islamismo e la cultura europea* (Florence, 1956); Norman Daniel, *Islam and the West: The Making of an Image* (Edinburgh, 1960); Ekkehart Rotter, *Abendland und Sarazenen* (Berlin, 1986); and Kenneth M. Setton, *Western Hostility to Islam and Prophecies of Turkish Doom* (Philadelphia, 1992). Premodern Muslim perceptions of Christian Europe have received far less attention. For some Arab views of Byzantium, see A. A. Vasiliev, *Byzance et les Arabes,* revised and expanded by H. Grégoire, M. Canard, and M. Nallino, 3 vols. (Brussels, 1935–1968), including translations and discussions of Arabic texts; and Ahmad Shboul, *Al-Mas'ūdī*

and His World: A Muslim Humanist and His Interest in Non-Muslims (London, 1979), of which chapter 6 (pp. 227–284) examines a major Arab historian's presentation of the country, capital, people, history, culture, and contemporary situation of the Byzantine Empire. Shboul devotes only a few pages (pp. 189–193) to western Europe, which in general received little attention from Muslim authors before the nineteenth century. On mutual perceptions during the Crusades, see Benjamin Z. Kedar, *Crusade and Mission: European Approaches Toward the Muslims* (Princeton, 1984); Claude Cahen, *Orient et Occident au temps des Croisades* (Paris, 1993); Amin Maalouf, *The Crusades Through Arab Eyes* (London, 1984); and Emmanuel Sivan, *L'Islam et la Croisade: Idéologie et propagande dans les réactions musulmanes aux Croisades* (Paris, 1968). For a study of what Muslims knew and thought about Europe, see Bernard Lewis, *The Muslim Discovery of Europe* (New York, 1982).

2. On Muslim knowledge of classical antiquity, see Franz Rosenthal, *Das Fortleben der Antike im Islam* (Zurich, 1965), English translation, *Classical Heritage in Islam* (Berkeley, 1975).

3. On Islamic tolerance, see Rudi Paret, "Toleranz und Intoleranz im Islam," *Saeculum* 21 (1976): 344–365; Francesco Gabrieli, "La tolleranza nell' Islam," *La cultura* 10 (1972): 257–266, reprinted in Gabrieli, *Arabeschi e studi islamici* (Naples, 1973), pp. 25–36; Adel Khoury, *Toleranz im Islam* (Munich, 1980); and Bernard Lewis, *The Jews of Islam* (Princeton, 1984), pp. 3–66.

4. Rashīd al-Dīn, *Histoire des Francs,* ed. and trans. K. Jahn (Leiden, 1951), text p. 11, translation p. 24.

5. Jacques-Paul Migne, ed., *Patrologiae cursus completus, series latina,* vol. 126, col. 816, quoted in Marius Canard, "La Guerre sainte dans le monde islamique et dans le monde chrétien," *Revue africaine* (Algiers) (1936): 605–623. See also Anouar Hatem, *Les Poésies épiques des Croisades* (Paris, 1932), pp. 36–40. For a discussion and comparison of Muslim and Christian notions of holy war and martyrdom in battle, see Albrecht Noth, *Heiliger Krieg und Heiliger*

Kampf in Islam und Christentum: Beiträge zur Vorgeschichte und Geschichte der Kreuzzüge (Bonn, 1966).

6. On this question, see the pioneer work of David Ayalon, *Gunpowder and Firearms in the Mamluk Kingdom: A Challenge to a Medieval Society* (London, 1956, 1978). For a general survey of the history of firearms in the Muslim world, see *EI²*, s.v. "Bārūd."

7. Quoted in Ayalon, *Gunpowder,* pp. 94–95.

8. Selaniki Mustafa Efendi, *Tarih-i Selâniki,* ed. Mehmet Ipşirli (Istanbul, 1989), vol. 1, p. 334.

9. Charles Issawi, "Technology, Energy, and Civilization: Some Historical Observations," *International Journal of Middle Eastern Studies* 23 (1991): 281–289.

10. Karl Jahn, "Das iranische Papiergeld," *Archiv Orientální* 10 (1938): 308–340.

11. On the history of printing in the Muslim world, see *EI²*, s.v. "Matbaʿa." On early Hebrew printing, see A. Freimann, *Über hebräische Inkunabeln* (Leipzig, 1902); and A. M. Habermann, *Toldot ha-Sefer ha-Ivri* (Jerusalem, 1945).

12. Kemal Atatürk, Speech delivered at Izmir, 31 January 1923, in *Atatürk'ün Söylev ve Demeçleri II (1906–1938)* (Ankara, 1952), p. 84. Atatürk returned to this theme on other occasions, notably in a speech delivered at Kastamonu on 30 August 1926 (ibid., p. 219).

13. On Muslim perceptions of Western women, see Lewis, *Muslim Discovery,* pp. 281, 286–293. On the portrayal of the Western woman in modern Arabic literature, see Rotraud Wielandt, *Das Bild der Europäer in der modernen arabischen Erzähl- und Theaterliteratur* (Beirut, 1980), especially pp. 489–553; and Wielandt, "Factors Determining the Picture of the European Woman in Modern Arabic Fiction," *Tarih: Papers in Near Eastern Studies* 2 (1992): 17–40.

14. Bernard Vincent, *1492: L'Année admirable* (Paris, 1991), pp. 72–78.

15. Ibid., p. 78.

16. A. Adnan [Adivar], *La Science chez les Turcs ottomans* (Paris, 1939), p. 57.

II

1. Paolo Preto, *Venezia e i Turchi* (Florence, 1975), p. 132.

2. On the image of the Prophet in the medieval and modern West, see *EI²*, s.v. "Muḥammad," pt. 3, where extensive bibliographical guidance is given. Among earlier literature, reference must still be made to the classical study by Alessandro d'Ancona, "La leggenda di Maometto in Occidente," *Giornale storico della letteratura italiana* 13 (1899): 199–281.

3. Already in the early eleventh century, the French chronicler Adhémar of Chabannes holds the Jews responsible for the destruction, by order of the Fatimid caliph al-Ḥākim, of the Church of the Holy Sepulchre in Jerusalem, and alleges that western Jews sent letters to the east accusing the Christians and warning that they were preparing to send an army of westerners against the eastern Saracens. See Adhémar of Chabannes, *Chronican Aquitanicum et Francicum* 137, cited in Hatem, *Les Poésies épiques,* p. 43. The claim by Allan Harris Cutler and Helen Elmquist Cutler, *The Jew as Ally of the Muslim: Medieval Roots of Anti-Semitism* (Notre Dame, 1986), that this perceived association is the root cause of anti-Semitism seems overstated.

4. Shlomo Eidelberg, ed. and trans., *The Jews and the Crusades: The Hebrew Chronicles of the First and Second Crusades* (Madison, 1977), p. 22.

5. On these, see Bernard Lewis, "The Pro-Islamic Jews," *Judaism* 17 (1968): 391–404, reprinted with revisions in Lewis, *Islam in History,* 2nd ed. (Chicago, 1993), pp. 137–151.

6. Johannes Hofer, *Johannes Kapistran: Ein Leben im Kampf um die Reform der Kirche* (Heidelberg, 1964), vol. 2, p. 155. Compare

H. H. Ben-Sasson, ed., *A History of the Jewish People* (Cambridge, Mass., 1976), p. 580.

7. On the Noahid commandments, see *Midrash Bereshit Rabba, Noah,* 34.8, English translation by H. Freedman and Maurice Simon, *The Midrash Rabbah: Genesis* (London, 1977), vol. 1, p. 272.

8. There is an extensive literature on the *marranos*. For a general survey of their fate in Spain and in the Spanish empire, see Antonio Domínguez Ortiz, *Los judeoconversos en España y América* (Madrid, 1978). On the quest for "purity of blood" in Spain, see Albert Sicroff, *Les Controverses des statuts de "Pureté de sang" en Espagne du XVe au XVIIe siècle* (Paris, 1960). For a remarkable study of the spiritual vicissitudes of a reconverted Iberian *marrano* in Ottoman Jerusalem, see Minna Rozen, *Jewish Identity and Society in the Seventeenth Century: Reflections on the Life and Work of Refael Mordekhai Malki* (Tübingen, 1992). On the practice of religious dissimulation by Shi'ite Muslims, see *EI*[1], s.v. "Taķīya."

9. A lecture delivered in Venezuela in 1985, inaugurating a series of "Sesiones Solemnes" that culminated in October 1992, does not actually celebrate the expulsion but justifies it, along with the Inquisition, as necessary measures for national unification and therefore as the lesser evil. See Jorge Olavarría, *Conmemoración del descubrimiento de América* (Caracas, 1985), especially pp. 13, 16–18. My thanks are due to Netanel Lorch for drawing my attention to this lecture.

10. For a wide-ranging survey, see *The Sephardi Heritage: Essays on the History and Cultural Contribution of the Jews of Spain and Portugal,* vol. 1: *The Jews in Spain and Portugal Before and After the Expulsion of 1492,* ed. R. D. Barnett (London, 1971); vol. 2: *The Western Sephardim,* ed. R. D. Barnett and W. M. Schwab (Grendon, Northants, 1989).

11. A famous case was that of the great Jewish theologian and physician Maimonides, who was born in Cordova and spent his early years there and in Morocco. According to one source, he was

obliged, under the rule of the fanatical Almohades, to make public profession of Islam, but at the first opportunity he and his family fled to the more tolerant atmosphere of Egypt, where he professed Judaism and became a leader of the Jewish community in Cairo. One day a Muslim visitor from Morocco claimed to recognize him, denounced him as an apostate from Islam, and demanded the full penalty of the law for this capital offense. The Egyptian qadi ruled that since the alleged conversion was not voluntary it was not valid, and the question of apostasy did not therefore arise. Maimonides was free to practice his religion and resume his career. The story is told by a thirteenth-century Egyptian scholar in his history of physicians. See Ibn al-Qiftī, *Ta'rīkh al-Ḥukamā'*, ed. J. Lippert (Leipzig, 1903), pp. 317–319.

12. See Attilio Milano, *Storia degli ebrei in Italia* (Turin, 1963), pp. 250–252, where earlier sources and studies are cited.

13. Eliyahu Capsali, *Seder Eliyah Zūṭā*, ed. Aryeh Shmuelevitz (Jerusalem, 1975), vol. 1, pp. 218–219, and an earlier, partial edition by M. Lattes, *De vita et scriptis Eliae Kapsalii* (Padua, 1869), pp. 12–13. On Capsali and his work, see Moritz Steinschneider, *Die Geschichtsliteratur der Juden* (Frankfurt, 1905), pp. 93–94. There are numerous other expressions of gratitude in Jewish sources for the haven offered by the Ottoman Empire in a time of persecutions in Christian Europe. One of them, known as the Edirne letter, appears to have been written in the early fifteenth century—that is, before the final expulsion from Spain, but at a time when the persecution was already well under way. The writer, who describes himself as a Jew of French origin born in Germany and living in Edirne, speaks eloquently of the torments endured by Jews in Europe and of the tolerance and tranquillity of Turkey. More than a century later, a Portuguese Jew called Samuel Usque wrote a famous book, *The Consolation for the Tribulations of Israel*, in which he offers both divine and human consolations to his coreligionists.

Among the latter, "the most signal is great Turkey." On these, see Lewis, *Jews of Islam,* pp. 135–136.

14. Capsali, *Seder Eliyah Zūṭā,* vol. 1, pp. 7–15.

15. On these documents, see Bernard Lewis, *Notes and Documents from the Turkish Archives: A Contribution to the History of the Jews in the Ottoman Empire* (Jerusalem, 1952), pp. 28–34, reprinted in Lewis, *Studies in Classical and Ottoman Islam (7th–16th Centuries)* (London, 1976); and, on the cancellation of the previous orders, Uriel Heyd, *Ottoman Documents on Palestine, 1552–1615* (Oxford, 1960), pp. 167–168. On these and other similar orders for resettlement, see Lewis, *Jews of Islam,* pp. 121–125.

16. On the *dhimma,* see *EI*² s.v., where sources and studies are cited.

17. On the *mudéjares,* see L. P. Harvey, *Islamic Spain, 1250–1500* (Chicago, 1990); and *EI*², s.v.

18. On this question, see Friez Meier, "Über die umschrittene Pflicht des Muslims, bei nichtmuslimischer Besetzung seines Landes auszuwandern," *Der Islam* 68 (1991): 65–86; Bernard Lewis, "Legal and Historical Reflections on the Position of Muslim Populations Under Non-Muslim Rule," *Journal of the Institute of Muslim Minority Affairs* 13 (1992): 1–16, reprinted in Lewis, *Islam and the West* (New York, 1993), pp. 43–57; and A. L. Udovitch, "Muslims and Jews in the World of Frederic II: Boundaries and Communication," *Princeton Papers in Near Eastern Studies* 2 (1993): 83–104.

19. Abu'l-'Abbās Aḥmad ibn Yaḥyā al-Wansharīsī, "Asnā al-matājir fi bayān aḥkām man ghalaba 'alā waṭanihi al-Naṣārā wa-lam yuhājir," ed. Ḥusayn Mu'nis, *Revista del Instituto Egipcio de Estudios Islámicos en Madrid* 5 (1957): 129–191.

20. On the capitulation of 1491, see Harvey, *Islamic Spain,* pp. 314–321.

21. On the *moriscos,* especially the document of expulsion, see

Mercedes García Arenal, *Los moriscos* (Madrid, 1975), pp. 251–255; Antonio Domínguez Ortiz and Bernard Vincent, *Historia de los moriscos; vida y tragedia de una minoría* (Madrid, 1978); Andrew C. Hess, "The Moriscos: An Ottoman Fifth Column in Sixteenth-Century Spain," *American Historical Review* 74 (1968): 1–25; and *EI²*, s.v. On Cisneros, see also Harvey, *Islamic Spain*, pp. 392–334. On the arrival and resettlement of Spanish Muslims and *moriscos* in North Africa, see J. D. Latham, "Towards a Study of Andalusian Immigration and Its Place in Tunisian History," *Cahiers de Tunisie* 5 (1957): 203–252; and Latham, "The 'Andalus' in North Africa," *EI²*, vol. 1, pp. 496–497.

22. For a full discussion, see Sicroff, *Les Controverses des statuts de "Pureté de sang" en Espagne;* see also S. W. Baron, *A Social and Religious History of the Jews,* 2nd ed. (New York, 1969), vol. 13, pp. 84–91. For a brief account, see Bernard Lewis, *Semites and Anti-Semites* (New York, 1986), pp. 82–84.

23. See M. Benady, "The Settlement of Jews in Gibraltar, 1704–1983," *Transactions of the Jewish Historical Society of England* 26 (1979): 87–110; Benady, "The Jewish Community of Gibraltar," in *Sephardi Heritage,* vol. 2: *Western Sephardim,* ed. Barnett and Schwab, pp. 144–179; Sir Joshua Hassan, *The Treaty of Utrecht and the Jews of Gibraltar* (London, 1970); and A. B. M. Serfaty, *The Jews of Gibraltar Under British Rule* (Gibraltar, 1933). I. López de Ayala denounces Britain for these breaches of the treaty, in *Historia de Gibraltar* (Madrid, 1792), p. 322.

24. On the absorption of Iberian Jewish exiles in North Africa, see H. Z. (J. W.) Hirschberg, *A History of the Jews in North Africa* (Leiden, 1974), vol. 1, pp. 362–446; in the east, see Minna Rozen, *In the Mediterranean Routes: The Jewish-Spanish Diaspora from the Sixteenth to Eighteenth Centuries* (in Hebrew) (Tel Aviv, 1993), esp. chap. 1; Mark Alan Epstein, *The Ottoman Jewish Communities and Their Role in the Fifteenth and Sixteenth Centuries* (Freiburg, 1980); Lewis, *Jews of Islam,* chap. 3; and Aryeh Shmuelevitz, *The Jews of the*

Ottoman Empire in the Late Fifteenth and the Sixteenth Centuries (Leiden, 1984). Haim Gerber, *Economic and Social Life of the Jews in the Ottoman Empire in the 16th and 17th Centuries* (in Hebrew) (Jerusalem, 1982), provides a valuable collection of contemporary Jewish and Turkish documents. On the Sephardic Jews in the Ottoman lands in later times, see Aron Rodrigue, ed., *Ottoman and Turkish Jewry: Community and Leadership* (Bloomington, 1992); Avigdor Levy, *The Sephardim in the Ottoman Empire* (Princeton, 1992); Stanford J. Shaw, *The Jews of the Ottoman Empire and the Turkish Republic* (New York, 1991); Walter E. Weiker, *Ottomans, Turks, and the Jewish Polity: A History of the Jews of Turkey* (New York, 1992); and Esther Benbassa and Aron Rodrigue, *Juifs des Balkans: Espaces judéo-ibériques XIVᵉ–XXᵉ siècles* (Paris, 1991). For some Turkish perceptions of the position and roles of Jews and other non-Muslims in the Ottoman Empire, see Ali Ihsan Bağış, *Osmanlı ticaretinde gayri müslimler: Kapitülasyonlar—beratlı tüccarlar, Avrupa ve hayriye tüccarları (1750–1839)* (Ankara, 1983); and Gülnihâl Bozkurt, *Alman-İngiliz belgelerinin ve siyasî gelişmelerin ışığı altında gayrimüslim Osmanlı vatandaşlarının hukukî durumu (1839–1914)* (Ankara, 1989).

25. His major work on the history and literature of Muslim Spain was entitled *Nafḥ al-Ṭīb min ghuṣn al-Andalus al-Ratīb* (The Perfumed Fragrance from the Verdant Branch of Andalus) and was written in Cairo in 1629. The work remained unknown until 1840, when the Spanish scholar Don Pascual de Gayangos published an English version adapted from the first part, dealing with the history of Muslim Spain. The Arabic text of this part was published for the first time in Leiden between 1855 and 1861 by a team of scholars—the Dutchman Reinhart Dozy, the Frenchman Gustave Dugat, the German Ludolf Krehl, and the Englishman William Wright. The complete Arabic text was published in Egypt in 1862 and in many subsequent editions.

26. On the revival of interest in Spanish Islam among Muslims,

see Henri Pérès, *L'Espagne vue par les voyageurs musulmans de 1610 à 1930* (Paris, 1937); and B. Lewis, "The Cult of Spain and the Turkish Romantics," in Lewis, *Islam in History,* pp. 129–133, originally published in French in *Études d'orientalisme dédiées à la mémoire de Lévi-Provençal* (Paris, 1962), vol. 2, pp. 185–190. The same French volume contains an important study by Aziz Ahmad, "Islam d'Espagne et Inde musulmane moderne," in *Études,* vol. 2, pp. 461–470. On the contribution of the Spanish Arabists, see Manuela Manzanares de Cirre, *Arabistas españolas del siglo XIX* (Madrid, 1972); and Victor Morales Lezcano, *Africanismo y orientalismo español en el siglo XIX* (Madrid, 1988).

III

1. Cited in David Abulafia, *Spain and 1492* (London, 1992), p. 69.

2. C. F. Beckingham, *The Achievements of Prester John* (London, 1966).

3. Otto von Freising, *Ottonis episcopi Frisingensis chronica, sive historia de duabus civitatibus,* ed. Adolf Hofmeister (Hanover-Leipzig, 1912), bk. 7, chap. 33, pp. 363–367, translated in Beckingham, *Achievements of Prester John,* p. 4. On the crusader background of Columbus's voyages, see Abbas Hamdani, "Columbus and the Recovery of Jerusalem," *Journal of the American Oriental Society* 99 (1979): 39–48.

4. Ibn Ḥajar al-'Asqalānī, *Inbā' al-ghumr bi-anbā' al-'umr,* ed. Ḥasan Ḥabashī (Cairo, 1972), vol. 3, pp. 426–427; Aḥmad ibn 'Alī al-Maqrīzī, *Kitāb al-Sulūk li-ma'rifat duwal al-mulūk,* ed. Said A. F. Ashour (Cairo, 1972), pt. 2, vol. 4, pp. 495–496; Abu'l-Maḥāsin ibn Taghrī-birdī, *Al-Nujūm al-zāhira fī mulūk Miṣr wa'l-Qāhira* XIV, ed. Jamāl Muḥammad Muhriz and Fahīm Muḥammad Shaltūt (Cairo, 1971), pp. 324–326.

5. Pliny, *Natural History*, 3.1.2–5. The seventeenth-century Ottoman polymath Ḥājji Khalīfa, also known as Kâtib Çelebi, gives a slightly different version, clearly derived from European sources, and adds: "They call the lands of Ethiopia and Egypt, which divide the Mediterranean from the Red Sea, Africa, and they call the New World America" (*Tuhfetül-Kibar fi Esfari'l-Bihar*, ed. Orhan Şaik Gökyay [Istanbul, 1973], p. 7). For a rather loose English translation, see Haji Khalifeh, *The History of the Maritime Wars of the Turks*, trans. James Mitchell (London, 1831), p. 4. Mitchell's translation was based on the first printed edition, published in Istanbul in 1729. It was the second Turkish book printed in Turkey, and contains many typographical errors, some of which are listed in an appendix of more than two hundred errata.

6. On the medieval Muslim view of the geography of the world, see André Miquel, *La Géographie humaine du monde musulman*, 4 vols. (Paris, 1967–1988); Ignatii Yulyanovič Kračkovskii, *Arabskaya Geografičeskaya literatura*, vol. 4 of *Izbranniye sočineniya* (Moscow, 1957), Arabic translation by Ṣalāḥ al-Dīn 'Uthmān Hāshim, *Ta'rīkh al-Adab al-Jughrāfī al-'Arabī*, 2 vols. (Cairo, 1963, 1965; Beirut, 1987).

7. On the New World map of 1513, see Svat Soucek, "Islamic Charting in the Mediterranean," in *The History of Cartography*, pt. 2, vol. 1: *Cartography in the Traditional Islamic and South Asian Societies*, ed. J. B. Harley and David Woodward (Chicago, 1992), pp. 269–272, where earlier studies are examined. For an excellent survey, see Andrew C. Hess, "Piri Reis and the Ottoman Response to the Voyages of Discovery," *Terrae Incognitae* 6 (1974): 19–37.

8. See Thomas D. Goodrich, *The Ottoman Turks and the New World: A Study of "Tarih-i Hind-i garbi" and Sixteenth Century Ottoman Americana* (Wiesbaden, 1990); and *Tarih-i Hind-i Garbi veya Hadis-i Nev* (A History of the Discovery of America) (Istanbul, 1987).

9. The text was discovered and edited by the Lebanese Jesuit

scholar Anṭūn Rabbāṭ and published as "Riḥlat awwal sā'iḥ sharqī ilā Amerika," *Al-Mashriq,* September–December 1905. For a brief description, see Abelardo Chediac, "Primer viaje de un oriental a la América," *América Española* 25 (1940): 87–98. A Moroccan ambassador to Spain in 1779 includes in his report what must surely be the first Muslim account of the American Revolution. See Muḥammad ibn 'Uthmān al-Miknāsī, *Al-Iksīr fī fikāk al-asīr,* ed. Muḥammad al-Fāsī (Rabat, 1965), p. 97. On the subsequent growth of knowledge, see Ami Ayalon, "The Arab Discovery of America in the Nineteenth Century," *Middle Eastern Studies* 20 (1984): 5–17.

Index

Arabic names are indexed in the form in which they are most familiar in the West. Generally, names from earlier periods are indexed under the given or first name, while more current persons are found under the last name.

Abolition, 72
Adhémar of Chabannes, 86n3
Adıvar, Adnan, 26
Aegean Sea, 66
Africa, 14, 37, 68, 72
  and Europe, 61, 62
  and European expansion, 7, 8, 70, 71, 73
  geographical classification of, 63–68, 70
  Islam in, viii, 10, 18, 67
  naming of, 65, 66–67
  North. *See* North Africa
  self-definition of, 64–65
  South. *See* South Africa
  viewed by Europeans, 64, 71
Africanism, 71
Albania, 26
Alfonso VI (king), 19
Algeria, 66. *See also* North Africa
Almohades, 87–88n11
America, viii, 9, 37, 72, 74, 77. *See also* United States
  conquest in, 8
  discovery of, viii, 5–6, 8, 69, 72–73
  and European expansion, 7, 8, 73
  geographical classification of, 65, 69, 70

America (*Cont.*)
as leader of the West, 74, 77
naming of, 6
pre-Columbian, 6, 9–10, 14, 64–65, 72
viewed by Islam, 14, 69–70
American Revolution, 93n9
Anatolia, 12, 20, 43. *See also* Turkey
Ancona, 38
Al-Andalus, 50
Anglo-Moroccan Treaty, 49
Apostasy, 38, 87–88n11
Arabian Peninsula, 58
Arabs, 7, 66, 70–71
Aragon, 47
Aristotle, 64
Armenia, 59
Arūfa, 70
Asia, viii, 9–10, 11, 17, 23, 37
and European expansion, 7, 8, 71, 73
geographical classification of, 63–68, 70
Islam in, viii, 10, 18, 67
naming of, 65, 66, 67
self-definition of, 64
Southeast, 58, 62
viewed by Europeans, 14, 64, 71
viewed by Muslims, 14
Asia Minor, 66
Asianism, 71
Atatürk, Kemal, 24
Averroës, 45
Aztecs, 7, 70

Balkan Peninsula, 10, 12, 50
Berberiscos, 50
Beyazid, Sultan, 39, 40
Bible, 15, 33–34, 39, 63, 69
Bilād al-Sūdān, 67
Black Sea, 7
Bohemond VI (prince), 61
Britain, 12, 48–49, 71. *See also* England; Europe
and Gibraltar, 48–50
Budapest, 41, 43
Buddhism, 11, 15
Byzantine Empire, 13–14, 18, 20, 44, 66, 83–84n1
and Muslims, 12, 13, 20

Capsali, Eliyahu, 39–40
Carthage, 66
Caspian Sea, 7
Castile, 47
Catholics, 26, 30, 32–33, 37, 51
Ceylon, 58
China, 9–10, 14, 58, 65, 66
and Europe, 7, 8–9
Islamic expansion into, 18
papermaking imported from, to Islamic world, 16
Christendom, 34. *See also* Christianity and the Christians; Europe; West
economic rivalry with Islam of, 58–59
holy war in, 19
medieval, 15–16, 17
second front against Islam of, 57–63
Christianity and the Christians, 10, 14, 31, 40, 66, 76. *See also* Catholics; Christendom; Protestantism
apostasy from, 38
Africa and Asia viewed by, 14
as European, 10
and the Iberian Peninsula, viii, 8, 17
Islam and Muslims viewed by, 11, 12–13, 14–15, 30, 31–32, 33, 34, 35, 45, 58, 68, 86n3
Judaism viewed in, 26, 30–32, 33–35, 36, 86n3
Nestorians, 59, 60–61
similarity with Islam of, 9–10, 14, 15, 66
universalist aims and claims of, 10, 11, 14, 15
viewed by Jews, 40
viewed in Islam, 10, 11, 14–15, 68, 70–71, 83–84n1
war against Islam of, 57–58
Cisneros, Francisco Jiménez de, 47
Cold war, 75
Columbus, 5, 8, 58, 69, 73
historical revision of, 6
Confucianism, 15
Conquest of Granada (Irving), 52
Conquistadores, 8
Consolation for the Tribulations of Israel (Usque), 88n13
Constantinople, 12. *See also* Istanbul

Turkish conquest of, 26, 33, 50
Crete, 39
Crusades and Crusaders, 10, 20, 30–31, 44, 68, 83–84n1
Cyprus, 42–43
Cyrus (king of Persia), 39

Damascus, 42, 43, 60–61
Dhimma, 44–45
Disraeli, Benjamin, 42
Domesday Book, 23
Donskoi, Dmitri, 19
Dozy, Reinhart, 91n25
Dugat, Gustave, 91n25

East Indies, 71
Edict of Expulsion of the Jews, 35–36
Edirne letter, 88n13
Egypt, 63, 65, 67, 76
    Christianity in, 11
    and Islamic expansion, 11, 21, 69
    Jewish exiles in, 87–88n11
England, 13, 22, 36, 37. *See also* Britain
Essai sur l'histoire des arabes et des mores d'Espagne (Viardot), 52
Ethiopia, 10, 59, 61
Eurocentrism, 65, 71, 73, 76
Europe, Western, 14, 66, 76. *See also* Christendom; West; *individual countries*
    and Africa, 7, 8, 64, 70, 71, 73
    and America, 5, 9, 69, 72–73
    and Asia, 64, 71, 73
    and China, 7, 8–9
    Christianity in, 20
    disunity of, as advantage, 22
    Eastern, 19
    economic relations with Islam of, 58–59, 62, 73
    entity defined, in Old World, 63–67
    expansion of, viii, 6–7, 8, 15, 25, 62, 70, 71, 73–74
    geographical classification by, 64, 69, 70–71
    and India, 7, 8–9, 71
    intercontinental lines of communication created by, 71–72
    Islam in, viii, 10, 11–12, 18, 59
    Islam viewed in, 8, 9, 12–13, 58, 67, 83–84n1

Jews in, 29, 30, 31, 37, 38, 44, 88n13
    medieval, 15–16
    and multiculturalism, 76
    Muslims viewed in, 8, 9, 12–13, 31–32, 58, 67, 68, 83–84n1
    recent changes in civilization in, 74–75
    religion and identity in, 69
    and Russia, 74
    second front against Islam sought by, 59–63
    slavery in, 72
    technological sophistication of, 20–23, 24
    viewed by Muslims, 9, 13–14, 70–71, 83–84n1
    voyages of discovery from, 53, 71–72, 73

Ferdinand of Aragon (king), 8, 36
Fertile Crescent, 76. *See also* Middle East; *individual countries*
Firearms, 20–21, 47. *See also* Gunpowder
Fondaco dei Turchi, 29
1492
    anniversaries of, vii, 24, 36–37, 87n9. *See also* America, discovery of; Granada, Christian conquest of; Jews, expulsion from Spain of
    Christian dominance in, viii
France, 19, 22, 71. *See also* Europe; West
    and the Jews in, 36, 37, 51
Franks, 17, 68, 70
Freeman, E. A., 32

Gama, Vasco da, 58, 62
Gautier, Theophile, 52
Gayangos, Don Pascual de, 91n25
Genoa, 5
Germans, 29, 51
Gibbon, Edward, 58
Gibraltar, 48–50, 57
Gladstone, William Ewart, 32
Gold, 62, 73
Golden Horde, 12
Grammática castellana (Nebrija), 25

Granada, 47
  Christian conquest of, vii, viii, 8,
    11, 19, 26, 32, 35, 36–37, 46, 73
  Jews expelled from, 36
  Muslims of, 46–47
Greece, 8, 15, 20, 51, 68–69
  geographical classification, 63–64,
    65, 66, 71
  and inheritance of Islam and Chris-
    tianity, 14
  Turkish conquest of, 26
  viewed by Muslims, 13, 68
Griegos, 50
Guilt, 73, 75
Gunpowder, 20, 23. *See also* Firearms

al-Ḥākim, 86n3
Hellas, 68
Hellenistic civilization, 16, 64
Heraclius (emperor), 18
Hernani (Hugo), 52
Hethoum (king), 61
Hijra, 45–46
Hinduism, 10, 15
*History of the Jews in Spain and Portu-
    gal* (Lindo), 52
Holland. *See* Netherlands
Holy War, 18–19. *See also* Crusades
    and Crusaders; *Jihād*
Hong Kong, 78
House of Islam, 14
House of War, 14
Hugo, Victor, 52
Hungary, 41
Huns, 7

Iberian Peninsula, 8, 36, 57, 62. *See
    also* Portugal and the Portuguese;
    Spain
  Christian reconquest of, 17, 19, 20
  Jews in, 17
  Muslims in, vii, viii, 7
  non-Christians in, vii, viii
Ibn Rushd, 45–46
Ibn Zunbul, 21
Iceland, 12
Imperialism, 20, 71, 79
Incas, 70
India, 10, 14, 58, 65, 66
  and Europe, 7, 8–9, 71
  and Islam, 16, 18

religions of, 10, 15, 65
Indian Ocean, 22, 67
Infidel, 68, 70
Inquisition, 47, 87n9
Iqlīm, 67
Iran, 17, 65. *See also* Persia
Irving, Washington, 52
Isabella of Castile (queen), 8, 36
Islam, 8, 14, 16–17, 30, 52, 65, 67,
    72. *See also* Middle East; Muslims
  America viewed in, 14, 69–70,
    93n9
  compared to other Old World civili-
    zations, 8–10
  decline of power of, 17–18
  diversity and pluralism in, 16–17
  economic relations with Christen-
    dom of, 58–59, 62, 73
  expansion of, 7, 10, 11–12, 18, 33,
    59
  and Jews, 37, 38
  medieval, 15–17
  Renaissance in, 25
  similarities with Christianity of, 9–
    10, 14, 15, 65
  Spanish, 51, 52
  universality of, 9–11, 15
  viewed by Christians, 11, 14–15,
    30, 31–32, 33, 35, 68. *See also*
    Muslims, viewed by Christians
  viewed in Europe, 8, 9, 12–13, 67,
    83–84n1
Issawi, Charles, 23
Istanbul, 37, 50. *See also* Constantino-
    ple
Italy, 51. *See also* Venice
  Islam in, 10, 11, 26
  and Jews, 37, 44
Ivan III (the Great) (tsar), 19

Japan, 65
Jengiz Khan, 6
Jews, 51. *See also* Judaism
  Christianity viewed by, 40
  in Europe, 29, 30, 31, 37, 38, 44,
    88n13
  expulsion from Portugal of, viii, 36,
    37, 38, 52–53
  expulsion from Spain of, vii, viii,
    34, 35–36, 37, 38–39, 44, 47,
    48, 52–53, 87n9, 88n13

forced conversion to Christianity of, 31, 34, 36, 37–38
linked with Muslims as enemies of the Church, 30–32, 86n3
persecuted during the Crusades, 30–31
in Spain, 17, 23, 26, 35
in Venice, 29, 30
viewed by Christians, 26, 30–32, 33–35, 36, 86n3
Jihād, 9, 12, 21, 63, 67
Christian response to, 18, 20
John VIII (pope), 19
Judaism, 14, 51–52. *See also* Jews

Karakhitay, 60
Katvan Steppe, Battle of, 60
Kitbuga Noyon, 60
Krehl, Ludolf, 91n25
Kulikovo, Battle of, 19

Language, 23, 78
Arabic, 17, 25, 67, 76
Castilian, 25
European diversity in, 17
Greek, 23, 44
Latin, 23, 25
Spanish, 25, 44
study of exotic, 76
vernacular standardized into literary, 25
of the West, studies by Asians and Africans, 70
Lebanon, 59, 60. *See also* Levant
Leo IV (pope), 19
Levant, 11, 12, 20. *See also* Lebanon; Syria
Levantini, 37
Libya, 63–64, 67
Lindo, Elias Haim, 52
Llull, Ramon, 57

Maimonides, 87–88n11
Mamluks, 21
al-Maqqarī, Ahmad ibn Muhammad, 52, 91n25
Marranos and marranism, 36, 37, 38, 47, 48
Mecca, 45
Medici, Lorenzo de', 24–25
Medina, 45

Mediterranean, 22, 31, 32–33, 65, 71
Messina, 19
Microasia. *See* Asia Minor
Middle East, 14, 18, 20, 66. *See also* Islam; *individual countries*
economic relations with Europe of, 62
printing rejected in, 23
Mills, 23
Minorca, 49
Missionaries, 8, 11
Mongols, 7, 12, 23, 59, 60–61
Moors, viii
viewed by Europeans, 12–13, 31, 32, 58, 68
Moriscos, 47–48, 89n21
Morocco, 22, 33, 49, 50, 67. *See also* Africa; North Africa
Islamic expansion into, 18, 49, 50
Mudéjar, 45, 48
Muhammad (Prophet), 10, 21, 29, 30, 45–46
Multiculturalism, 76–77
Muscovites, 7. *See also* Russia
Muslims. *See also* Islam
under Christian rule, 45–47
Christians and Christianity viewed by, 10, 11, 14–15, 68, 70–71, 83–84n1
ethnic terms used for, by Christians, 68
Europe viewed by, 9, 13–14, 70–71, 83–84n1
expulsion from Portugal of, viii
expulsion from Spain of, viii, 44, 47, 48
interest in America of, 14, 69–70, 93n9
linked with Jews as enemies of the Church, 30–32, 86n3
in Spain, 35, 44–45
trade with Europe of, 72
viewed by Christians, 11, 12–13, 14–15, 30, 31–32, 33, 34, 35, 45, 58, 68, 86n3
viewed in Europe, 8, 9, 12–13, 31–32, 58, 67, 68, 83–84n1

Naples, 36
Natural History (Pliny), 63
Navarre, 47

Navas de Tolosa, Las, 19
Nazis, 51
Nebrija, Antonio de, 25
Nestorians, 59, 60–61
Netherlands, 22, 37, 71
New World, 20, 72, 77
Nicephorus Phocas (emperor), 18
Noahid commandments, 86n7
Normans, 19
North Africa, 17. *See also* Africa; Algeria; Morocco; Tunisia
  Arabic language in, 17
  Christianity in, 11,12
  and European expansion, 71
  and Islamic expansion, 11, 33
  Jewish exiles from Spain in, 50
  Muslims in, 45, 47, 50

O'Connor, T. P., 32
Old Testament, 33–34, 39
Ostia, 19
Otranto, 26
Otto (bishop of Freising), 59–60
Ottoman Empire, 21, 23, 44, 48, 70.
  See also Turkey; Turks
  expansion of, 7, 12, 26, 33, 41
  Jewish exiles in, 37, 38, 41–44, 50–51, 88n13

Palestine, 11
Paper, 16, 23
Persia, 18, 23, 58, 66. *See also* Iran
  and Europe, 59, 71
Philippines, 58
Phoenicians, 64
Pliny, 63
Ponentini, 37
Portugal and the Portuguese, viii, 22, 33, 62, 65. *See also* Christendom; Europe; Iberian Peninsula; West
  Christian reconquest of, 17, 57
  Jews in, 38, 52. *See also* Jews, expulsion from Portugal of
  and Muslims, viii, 11, 21, 58
Prester John, 59–60, 73
Printing, 23, 24
Protestantism, 29, 30

Qur'ān, 69

Rabbāt, Antūn, 93n9
Racism, 79
Rashīd al-Dīn, 17
Reconquista, 19–20, 44–45, 57
Red Sea, 22, 67
Reformation, 26, 30
Reis, Piri, 69
Renaissance, 24, 25–26, 68
Rhineland, 44
Roman Empire, 8, 14, 15, 58, 68
  geographical classification by, 63–64, 65, 66–67
  Islamic expansion into, 11, 19
Russia, 7, 62, 71, 74. *See also* Soviet Union
  Christian reconquest in, 12, 20
  Islam in, viii, 10, 12, 19, 57, 62

Safed, 42–43
Sahara, 65
St. John of Capistrano, 31, 34–35
Salonika, 43, 50, 51
Sanjar, Sultan, 60
Santa Maria, Gonzalo Garcia de, 25
Saracens, 12–13, 68, 86n3
Scipio, 66
Seafaring, 21–22, 72
Selaniki Mustafa, 22
Selim I (sultan), 21, 69
Seljuks, 12, 60
Sepharad, 50
Sephardic Jews, 41, 50, 51, 52
Sexism, 79
Shi'ites, 59
Siberia, 7
Sicily, 12
Silver, 73
Slavery, 13, 24
  colonial, 72
Slavs, 68
South Africa, 7
Soviet Union, 75. *See also* Russia
Spain, viii, 5, 22, 58, 62. *See also* Christendom; Europe; Iberian Peninsula; West
  and Britain, 48–49
  Christian conquest of, 8, 12, 17, 19, 35, 44, 45, 46, 57
  and danger of Muslim counterattack, 32–33

Islam in, viii, 8, 10, 11, 12, 35, 44–47, 52
Jews in, 26, 33, 35, 45, 52. *See also* Jews, expulsion from Spain
of voyages of discovery from, viii, 53
Spices, 58, 62, 73
Strait of Gibralter, 57, 62, 63
Sudan, 67
Sugar, 62
Süleyman the Magnificent (sultan), 23, 51
Syria, 11, 18, 19, 66. *See also* Levant

Tales of the Alhambra (Irving), 52
Tarih-i Hind-i Garbi, 69–70
Tartary, viii, 57
Tatars
  in Russia, viii, 12, 19, 57, 62
  viewed by Europeans, 12–13, 68
Thessaloniki. *See* Salonika
Three Kings, Battle of, 33
Toledo, 19, 47
Tolerance and intolerance, 16–17, 30–31, 45–46, 84n3, 88n13
Traders, 8, 11
Treaty of Utrecht, 49
Tunisia, 66
Turkey, 24, 32, 51. *See also* Anatolia; Ottoman Empire
  Jewish exiles in, 23, 37, 38–44, 50, 88n13
Turks, 12, 20, 66, 70. *See also* Ottoman Empire
  Europe viewed by, 70–71
  and Jews, 31–32, 41–42
  and Venice, 26, 29–30
  and Vienna, 11, 26
  viewed by Europeans, 12–13, 31–32, 68

United States, 74. *See also* America; West
Usque, Samuel, 88n13

Valencia, 47
Venice, 26, 29–30, 37, 39, 43. *See also* Italy
Vespucci, Amerigo, 6
Viardot, Louis, 52
Vienna, 11, 12, 26, 34
Vietnam, 78
Vincent, Bernard, 25
Voyage en Espagne (Gautier), 52
Voyages of discovery, 53, 71–73, 74

al-Wansharīsī, Ahmad, 46
Warfare
  Christian, 18–20
  Muslim, 18. See also *Jihād*
Wars of Religion, 17
Weaponry, 18, 20–21, 24. *See also* Firearms; Gunpowder; Seafaring
West, 19, 24, 72, 75, 76–77. *See also* Britain; Christendom; Europe; France; United States
  defined, 74–75
  doubt in, 75–76
  foreign policy of, 24–25
  guilt in, 73, 75, 77
  imperialism of, 20
  rise of, 17–18
  study of other cultures in, 76–79
  technological sophistication of, 20–22, 24
West Indies, 70
Women, 24, 70 85n12
Wright, William, 91n25

Zimbabwe, 65